Lecture Notes in Mechanical Engineering

Series Editors

Fakher Chaari, National School of Engineers, University of Sfax, Sfax, Tunisia

Mohamed Haddar, National School of Engineers of Sfax (ENIS), Sfax, Tunisia

Young W. Kwon, Department of Manufacturing Engineering and Aerospace Engineering, Graduate School of Engineering and Applied Science, Monterey, CA, USA

Francesco Gherardini, Dipartimento Di Ingegneria, Università Di Modena E Reggio Emilia, Modena, Modena, Italy

Vitalii Ivanov, Department of Manufacturing Engineering Machine and Tools, Sumy State University, Sumy, Ukraine

Francisco Cavas-Martínez, Departamento de Estructuras, Universidad Politécnica de Cartagena, Cartagena, Murcia, Spain

Justyna Trojanowska, Poznan University of Technology, Poznan, Poland

Lecture Notes in Mechanical Engineering (LNME) publishes the latest developments in Mechanical Engineering - quickly, informally and with high quality. Original research reported in proceedings and post-proceedings represents the core of LNME. Volumes published in LNME embrace all aspects, subfields and new challenges of mechanical engineering. Topics in the series include:

- Engineering Design
- Machinery and Machine Elements
- Mechanical Structures and Stress Analysis
- Automotive Engineering
- Engine Technology
- Aerospace Technology and Astronautics
- Nanotechnology and Microengineering
- Control, Robotics, Mechatronics
- MEMS
- Theoretical and Applied Mechanics
- Dynamical Systems, Control
- Fluid Mechanics
- Engineering Thermodynamics, Heat and Mass Transfer
- Manufacturing
- Precision Engineering, Instrumentation, Measurement
- Materials Engineering
- Tribology and Surface Technology

To submit a proposal or request further information, please contact the Springer Editor of your location:

China: Dr. Mengchu Huang at mengchu.huang@springer.com
India: Priya Vyas at priya.vyas@springer.com
Rest of Asia, Australia, New Zealand: Swati Meherishi at swati.meherishi@springer.com
All other countries: Dr. Leontina Di Cecco at Leontina.dicecco@springer.com

To submit a proposal for a monograph, please check our Springer Tracts in Mechanical Engineering at http://www.springer.com/series/11693 or contact Leontina.dicecco@springer.com

Indexed by SCOPUS. The books of the series are submitted for indexing to Web of Science.

More information about this series at http://www.springer.com/series/11236

Grzegorz M. Królczyk ·
Piotr Niesłony · Jolanta Królczyk
Editors

Industrial Measurements in Machining

Editors
Grzegorz M. Królczyk
Politechnika Opolska
Opole, Poland

Piotr Niesłony
Politechnika Opolska
Opole, Poland

Jolanta Królczyk
Politechnika Opolska
Opole, Poland

ISSN 2195-4356 ISSN 2195-4364 (electronic)
Lecture Notes in Mechanical Engineering
ISBN 978-3-030-49909-9 ISBN 978-3-030-49910-5 (eBook)
https://doi.org/10.1007/978-3-030-49910-5

This Springer imprint is published by the registered company Springer Nature Switzerland AG
The registered company address is: Gewerbestrasse 11, 6330 Cham, Switzerland

Preface

This book gathers selected, rigorously peer-reviewed papers relating to recent problems in machining and grinding processes. The papers were presented at two conferences on the topic of machining and abrasive machining, held in Opole, Poland, on September 11–12, 2019.

The dynamic growth of a modern industry following the industrial revolution 4.0 has resulted in a growing demand for advanced machine parts made of construction materials with excellent operational properties. Nevertheless, the improved features of these parts, such as high surface quality, complex shape and excellent mechanical properties, have imposed the application of modern manufacturing processes taking into account the technological, economic and metrologic constraints. The two conferences have provided an important forum for researchers and industry experts to exchange knowledge regarding:

- novel production and machining techniques;
- optimization methods in manufacturing;
- phenomena occurring during manufacturing processes;
- novel measurement and characterization techniques.

We would like to express our gratitude to all conference members, including the scientific committee and our reviewers, who spared their valuable time to improve the quality, accuracy and importance of the papers selected for this book. We hope that the extensive information gathered here will foster the efficient selection of manufacturing methods for industrial applications, as well as the improvement of conditions, strategies and types of tools used during the machining and grinding processes, thus allowing advances in manufacturing performance and economics.

Contents

Mathematical Modelling of Core Roughness Depth During Hard Turning

Mite Tomov[1]([⊠]), Pawel Karolczak[2], Hubert Skowronek[2], Piotr Cichosz[2], and Mikolaj Kuzinovski[1]

[1] Faculty of Mechanical Engineering, "Ss. Cyril and Methodius" University in Skopje, Skopje, North Macedonia
mite.tomov@mf.edu.mk

[2] Department of Machine Tools and Mechanical Engineering Technologies, Faculty of Mechanical Engineering, Wroclaw University of Technology, Wroclaw, Poland

Abstract. This paper presents the mathematical model for modelling and predicting the core roughness depth (Rk parameter) during hard turning with ceramic inserts, using nonlinear function shape and a four-factorial plan experiments of the first order. The independent mathematical modelling input parameters comprise cutting speed (v), feed (f), tool nose radius (r_ε) and depth of cut (a_p), with repetition in the middle point of the investigated hyperspace. The research was carried out on working rings made of the steel C 55 (DIN), thermally enhanced on a 52 $\pm$ 2 HRC, while the processing was done on the CNC lathe model OKUMA LB 15-II (C-1S). The measurement of working rings, obtaining the roughness profile and calculation of the Rk parameters is done using the instrument Surtronic 3+ and the Talyprofile software. The results of the research are presented by nonlinear mathematical models for the Rk parameter and coefficient K_{Rk} obtained from the ratio between Rk and the sum of Rpk, Rk and Rvk. According to the obtained mathematical models, 3D graphs of the dependences between the Rk parameter, coefficient K_{Rk} and the input independent variables are also shown. The obtained mathematical models are explained, and appropriate conclusions are given.

Keywords: Mathematical modeling · Hard turning · Rk parameters

1 Introduction

It has been generally accepted that "hard turning" refers to turning of steels with hardness greater than 45 HRC. Many researchers study the physical phenomena and the technological effects arising from hard turning. Out of all technological effects occurring on the surface during hard turning, researchers pay mostly to the surface roughness and the attempts to model and predict it. The driving force behind this refers to the notion of replacing grinding with hard turning [1, 2].

Bartarya and Choudhury [1] investigate the cutting parameters effect on the cutting force and the surface roughness during finishing hard turning AISI52100 grade steel and develop the force and surface roughness regression models using full factorial design

© Springer Nature Switzerland AG 2020
G. M. Królczyk et al. (Eds.): IMM 2019, LNME, pp. 1–9, 2020.
https://doi.org/10.1007/978-3-030-49910-5_1

of experiments. Mohruni et al. [2] propose empirical mathematical models for surface roughness prediction depending on two hard turning input factors (cutting speed and feed rate) on AISI D2 steel using Cubic boron nitride insert. They conclude that the cutting speed and feed rate significantly affected the quality of machined surfaces. Furthermore, the surface roughness value reduced (became smoother) by increasing the cutting speed. On the other hand, as the feed rate increased, the surface roughness value increased significantly. Mital and Mehta [3] have conducted an analysis of developed surface prediction models and factors influencing the surface roughness. They developed the surface finish models for aluminum alloy 390, ductile cast iron, medium carbon leaded steel, medium carbon alloy steel 4130, and Inconel 718 for a wide range of machining conditions defined by cutting speed, feed and tool nose radius. The experimental data statistical analysis indicated that the type of the metal, speed and feed and tool nose radius strongly influence the surface finish. Unlike the effect of the cutting speed, the effects of feed and the tool nose radius on the surface finish were generally consistent for all materials. Özel et al. [4] used a four-factor, two-level factorial design with 16 replications to determine the effects of the cutting tool edge geometry, the feed rate workpiece hardness and the cutting speed on the resultant forces and surface roughness in finishing hard turning of AISI H13 steel using CBN tools. They reported that the cutting edge geometry had a significant effect on the surface roughness. Honed edge geometry and lower workpiece surface hardness resulted in better surface roughness. They also reported that the workpiece hardness, cutting edge geometry and cutting speed significantly influenced the force component. Sundram and Lambert [5, 6] developed the mathematical models for surface roughness prediction of AISI 4140 steel during the fine turning operation using both TiC coated, and uncoated tungsten carbide throw away tools. The parameters considered included speed, feed, cut depth, cut time, nose radius and tool type. Agrawal et al. [7] predict surface roughness using three regression techniques – multiple regression, random forest, and quantile regression. They concluded that the random forest regression model represents a superior choice over multiple regression models for predicting surface roughness during machining of AISI 4340 steel (69 HRC).

Generally, when the research talks about surface roughness prediction, it refers to the arithmetical mean deviation Ra. Figure 1, shows two roughness profiles obtained with cylindrical grinding and longitudinal turning with approximately equal values for the arithmetical mean deviation Ra. The shapes of the surface roughness profiles clearly show that these two profiles will behave differently during explanation even though they had almost identical values of the Ra parameter.

On the other hand, Maurici et al. [8], using artificial neural networks, propose models for parameters Rk, Rpk and Rvk for rough honing process. These parameters directly correlate with the Abbott-Firestone (bearing area) curve and are also known as function parameters. Grzesik [9] uses these parameters to examine the influence of tool wear on surface roughness in hard turning using differently shaped ceramic tools. By changing the shape of the bearing curves, he determines the impact of the wearing of the cutting tool on the change of the shape of the surface roughness profile. Grzegorz et al. [10] use the Rk, Rpk and Rvk parameters to compare the surface roughness of machined and of fused deposition modelled parts.

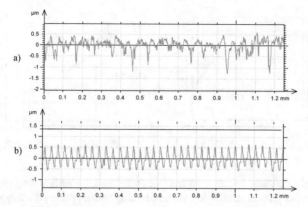

Fig. 1. Surface roughness profiles obtained by cylindrical grinding (a) and longitudinal turning (b), with approximately equal values for the arithmetical mean deviation $Ra \approx 0.2$ μm

This means that the Rk, Rpk and Rvk parameters provide greater detail in describing the surface roughness profile from a functional and explanation point of view. Hence, the authors of this paper decided to develop a mathematical model for predicting the core roughness depth, the Rk parameter, and a suitable correlation between Rk, Rpk and Rvk expresses by the newly introduced coefficient K_{Rk} during hard turning. Section 2 of this paper provides more details regarding the Rk parameter and the K_{Rk} coefficient.

2 Rk Parameter and Coefficient K_{Rk}

Rk parameter is standardized in ISO 13565-2:1996 [11], and graphically shown on Fig. 2. Figure 2 suggests that Rk parameter directly relate to the shape of the material ratio curve of the roughness profile. The material ratio curve of the roughness profile provides information regarding the percent of material ratio of the roughness profile to a certain depth and the shape of the material ratio curve directly depends on the shape of the roughness profile. Any changes to the shape of the material ratio curve directly impacts the values of the parameters Rk, Rpk and Rvk. Figure 3 shows material ratio curve and the Rk, Rpk and Rvk parameters, for surface roughness profile from Fig. 1.

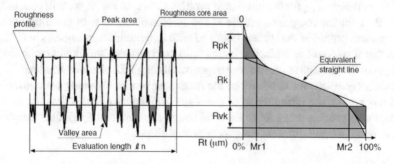

Fig. 2. Rk, Rpk and Rvk parameters [12].

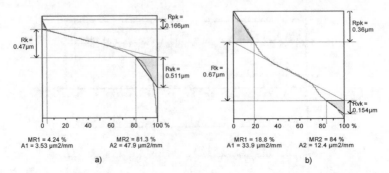

Fig. 3. Material ratio curve and *Rk, Rpk* and *Rvk* parameters, for surface roughness profile from Fig. 1, (a) cylindrical grinding and (b) longitudinal turning, obtained using the Talyprofile software.

Figure 3 clearly shows that the shape of the roughness profile material ratio curve of the surface obtained during grinding drastically differs from the shape of the roughness profile material ratio curve of the surface obtained during longitudinal turning. The values of the *Rk* parameter also reflects this difference, i.e. *Rk* = 0.47 μm for the roughness profile obtained from grinding, and for the roughness profile obtained from longitudinal turning *Rk* = 0.67 μm.

We propose the introduction of a coefficient K_{Rk} in order to obtain information about the changes to the shape of the roughness profile obtained from longitudinal turning as a function of the change of the cutting speed, feed, tool nose radius and depth of cut in the investigated hyperspace. We know that we can divide the height of the roughness profile into three parts. The core of the roughness profile, whose depth is defined by the Rk parameter, the protruding peaks whose height is defined by the Rpk parameter and protruding troughs whose depth is defined by the *Rvk* parameter. Any change of the shape of the roughness profile changes the shape of the roughness profile material ratio curve, thus directly impacting the values of the considered *Rk, Rpk* and *Rvk* parameters and the K_{Rk} coefficient.

We propose to determine the K_{Rk} coefficient as follows:

$$K_{Rk} = \frac{Rk}{Rpk + Rk + Rvk} \tag{1}$$

If we calculate K_{Rk} for the roughness profiles shown on Fig. 1, we will conclude that K_{Rk} = 0.41 for the roughness profile of the surface obtained with grinding, while for the roughness profile of the surface obtained with longitudinal turning K_{Rk} = 0.57. This means that if we want to replace grinding with hard turning, we should make sure that the depth of the roughness profile core expressed through Rk obtained from hard turning approximately equals the depth Rk of the roughness profile obtaining from grinding.

In this regard, the introduction of the K_{Rk} coefficient facilitates the comparison of the roughness profiles using *Rk* in the total height (*Rvk, Rk* and *Rpk*), thus avoiding the cases of equal *Rk* and different roughness profile shapes.

3   Experimental Research

The research mathematically models the Rk parameters using the steel C 55 (DIN). The material was prepared in the form of working rings, thermally enhanced on a 52 ± 2 HRC. The processing was done using a CNC lathe model OKUMA LB 15-II (C-1S) with a variable spindle speed 38–3800 rpm, feed rate 0.001–1000 mm/rev and 15 kW spindle drive motor. For the processing we used a cutting tool holder CSRNR 25×25 M12H3 (HERTEL) with $\kappa = 75°$; $\kappa_1 = 15°$; $\gamma = -6°$; $\lambda = -6°$; and ceramic cutting inserts HC2 from the producer NTK-Cutting tools. In order to avoid the cutting tool wearing effect, during processing we always used a new cutting edge.

We measured the working rings, obtained the roughness profile and calculated the Rk parameters using the instrument Surtronic 3+ and the Talyprofile software. As a λ_c profile filter we used a Gaussian filter with an appropriate sampling length. We used two sampling lengths of 0.25 mm and 0.8 mm depending on the value of RS_m.

The mathematical modeling of the Rk parameter and the K_{Rk} coefficient will make use of the nonlinear function shape and a four-factorial plan experiments from the first order. The independent mathematical modeling input parameters comprise cutting speed (v), feed (f), tool nose radius (r_ε) and depth of cut (a_p), with repetition in the middle point of the investigated hyperspace. Table 1 shows the values of the independent input parameters.

Table 1.   Input parameters and their levels

Sr. no.	Parameters	Level 1 (low)	Level 2 (central point)	Level 3 (high)
1	Cutting speed (v), m/min	67	94	133
2	Feed (f), mm/rev	0.1	0.17	0.3
3	Depth of cut (a_p), mm	0.4	0.56	0.8
4	Tool nose radius (r_ε), mm	0.8	1.2	1.6

This research made $2^4 = 16$ experiments, as well as four more as repetitions in the middle point, for a total of $2^4 = 16 + 4 = 20$ experiments. Table 2 presents the detailed plan of experiments, as well as the Rk and K_{Rk} values. These values (for Rk and K_{Rk}) are average values obtained from three measurements of the working rings, distributed at $120°$ angles along the entire circumference of the rings.

The mathematical modeling made use of the CADEX (Computer Aided Design and Analysis of Experiments) software of the Faculty of Mechanical Engineering – Skopje. To verify the adequacy of the obtained mathematical models we used a Fisher test with a significance level of $\alpha = 0.05$. The mathematical model accuracy was defined at a confidence interval of 95%.

Table 2. Experimentation and measured responses

No.	V (m/min)	f (mm/rev)	a_p (mm)	r_ε (mm)	Rk (μm)	K_{Rk}
1	67	0.1	0.4	0.8	1.170	0.54
2	133	0.1	0.4	0.8	1.037	0.48
3	67	0.3	0.4	0.8	5.450	0.39
4	133	0.3	0.4	0.8	5.717	0.43
5	67	0.1	0.4	1.6	1.197	0.42
6	133	0.1	0.4	1.6	1.490	0.68
7	67	0.3	0.4	1.6	7.200	0.46
8	133	0.3	0.4	1.6	4.620	0.34
9	67	0.1	0.8	0.8	0.657	0.44
10	133	0.1	0.8	0.8	0.775	0.64
11	67	0.3	0.8	0.8	4.263	0.58
12	133	0.3	0.8	0.8	3.707	0.62
13	67	0.1	0.8	1.6	0.704	0.53
14	133	0.1	0.8	1.6	0.816	0.6
15	67	0.3	0.8	1.6	3.317	0.43
16	133	0.3	0.8	1.6	5.247	0.66
17	94	0.17	0.56	1.2	1.627	0.42
18	94	0.17	0.56	1.2	2.613	0.53
19	94	0.17	0.56	1.2	2.440	0.53
20	94	0.17	0.56	1.2	2.217	0.47

4   Results and Discussion

After the mathematical processing of data for the Rk and K_{Rk} values we obtained the following mathematical models:

$$Rk = 21.748 \cdot v^{0.0609068} \cdot f^{1.418} \cdot a_p^{0.1204366} \cdot r_\varepsilon^{-0.6001313} \tag{2}$$

$$K_{Rk} = 0.1553540 \cdot v^{0.210255} \cdot f^{-0.0987966} \cdot a_p^{-0.0158564} \cdot r_\varepsilon^{0.2793446} \tag{3}$$

In the mathematical model for the Rk parameter, Eq. 2, the significant independent variables include the feed (f) and the tool nose radius (r_ε), while the insignificant ones include cutting speed (v) and depth of cut (a). This means that an increase in the feed (f) leads to an increase of the value of the Rk parameter and an increase of the radius (r_ε) reduces the value of the Rk parameter. The feed (f) has a much greater effect than the radius (r_ε).

In the mathematical model for the K_{Rk} coefficient, Eq. 3, the significant independent variables include only the tool nose radius (r_ε). Although the value of the exponent for the cutting speed (v) comes close to the value of the exponent for the tool nose radius

(r_ε), still mathematically the cutting speed (v) is not a significant input variable in this mathematical model. Although the significant input variable here is the tool nose radius, still the value of the exponent is small. The small value of the exponent for the tool nose radius (r_ε) approximately equal to the value of the exponent for the cutting speed (v), suggests that, generally, the shape of the roughness profile obtained from hard turning does not significantly change in the investigated hyperspace. Interestingly, the exponent of the tool nose radius (r_ε) has a positive sign, as opposed to the negative sign in Eq. (2). This suggests that an increase in the tool nose radius (r_ε) leads to an increase in the K_{Rk} coefficient, i.e. the share of the Rk parameter. This means that an increase in the tool nose radius, relative to a several times smaller feed, leads to a loss of the strictly deterministic character of the roughness profile. This also applies to the influence of the cutting speed, whose increase (of the cutting speed) can lead to vibrations during the cutting process and a disruption of the strictly deterministic character of the roughness profile.

Figures 4 and 5 also show 3D graphs of the dependences between the Rk parameter and K_{Rk} and the independent input variables.

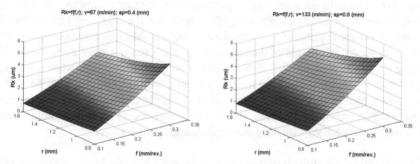

Fig. 4. 3D graphs of the dependences between the Rk parameter and the independent input variables (the feed (f) and tool nose radius (r_ε)).

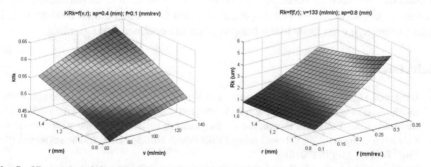

Fig. 5. 3D graphs of the dependences between the K_{Rk} parameter and the independent input variables (the cutting speed (v) and tool nose radius (r_ε)).

5 Conclusion

This research and the mathematical models for the Rk parameter and the K_{Rk} showed that:

- The significant input variables for the modelling of the Rk parameter include the feed (f) and the tool nose radius (r_ε), while cutting speed (v) and depth of cut (a) constitute the insignificant variables. The Rk parameter directly correlates with the feed, i.e. an increase in the feed will produce an increase of the Rk parameter On the other hand, an increase in the tool nose radius (r_ε) produces a reduction in the Rk parameter, i.e. the former inversely correlates to the latter. The feed has a greater impact than the tool nose radius, reflected in the values of the respective exponents. If we analyze the well-known theoretical equation for the maximum height of the roughness profile $Rt = f^2/8r_\varepsilon$ we will conclude that the above, regarding Rk applies here too. This means that by nature, Rk behaves identically as the surface roughness profile height parameters.
- The K_{Rk} coefficient model, the Eq. 3, and the exponents of the input variables suggest that there does not exist a dominant input parameter (cutting parameter) impacting the value of the K_{Rk} coefficient. Physically, this means that the roughness provide does not change its shape significantly as a result of changes to the input parameters. The share of Rk in the overall height of the roughness profile does not change drastically. The same applies to Rpk and Rvk as well. This conclusion shows that, when attempting to replace grinding with hard turning, one must consider the shape of the surface roughness profile. Hard turning generates a deterministic (periodic) roughness profile shape while grinding generates a stochastic (non-periodic) roughness profile shape, i.e. roughness profiles with different exploitation characteristics.

    In the future, the authors of this research propose to extend this research to other hard materials, using various cutting inserts for processing hard materials, in order to confirm and improve the results obtained from this research and make them more precise. In addition, we need to think how to reduce the value of the Rk parameter, for an unchanged height of the roughness profile obtained from longitudinal hard turning, in order to converge even more to the roughness profile obtained from grinding.

References

1. Bartaryaa, G., Choudhuryb, S.K.: Effect of cutting parameters on cutting force and surface roughness during finish hard turning AISI52100 grade steel. Procedia CIRP **1**, 651–656 (2012)
2. Mohruni, A.S., Yanis, M., Edwin, K.: Development of surface roughness prediction model for hard turning on AISI D2 steel using Cubic Boron Nitride insert. Jurnal Teknologi (Sci. Eng.) **80**(1), 173–178 (2018)
3. Mittal, A., Mehta, M.: Surface finish prediction models for fine turning. Int. J. Prod. Res. **26**(12), 1861–1876 (1988)
4. Özel, T., Hsu, T.K., Zeren, E.: Effects of cutting edge geometry, workpiece hardness, feed rate and cutting speed on surface roughness and forces in finish turning of hardened AISI H13 steel. Int. J. Adv. Manuf. Technol. **25**(3–4), 262–269 (2005). https://doi.org/10.1007/s00170-003-1878-5

5. Sundaram, R.M., Lambert, B.K.: Mathematical models to predict surface finish in fine turning of steel, Part 1. Int. J. Prod. Res. **19**(5), 547–556 (1981)
6. Sundaram, R.M., Lambert, B.K.: Mathematical models to predict surface finish in fine turning of steel, Part 2. Int. J. Prod. Res. **19**(5), 557–564 (1981)
7. Agrawal, A., Goel, S., Rashid, W.B., Price, M.: Prediction of surface roughness during hard turning of AISI 4340 steel (69 HRC). Appl. Soft Comput. **30**, 279–286 (2015)
8. Maurici, S.A., Irene, B.C., Xavier, L.P.: Neural network modelling of Abbott-Firestone roughness parameters in honing processes. Int. J. Surf. Sci. Eng. **11**(6), 512–530 (2017)
9. Grzesik, W.: Influence of tool wear on surface roughness in hard turning using differently shaped ceramic tools. Wear **265**(3–4), 327–335 (2008)
10. Grzegorz, K., Pero, R., Stanislaw, L.: Experimental analysis of surface roughness and surface texture of machined and fused deposition modelled parts. Tehnički vjesnik **21**(1), 217–221 (2014)
11. ISO 13565-2:1996, Geometrical Product Specifications (GPS) - Surface texture: Profile method; Surfaces having stratified functional properties - Part 2: Height characterization using the linear material ratio curve
12. Surface Texture, Contour Measuring Instruments. Explanation of Surface Characteristic Standards. https://inspectionengineering.com/uploads/2018/03/SurfaceFinishExplain

Unified Mathematic and Geometric Model of Cutting Edge Separation with Corner Radius in Turning

Storch Borys and Żurawski Łukasz[⊠]

Faculty of Mechanical Engineering, Department of Engineering of Technical and Information Systems, Technical University of Koszalin, ul. Racławicka 15-17, 75-620 Koszalin, Poland
lukasz.zurawski@tu.koszalin.pl

Abstract. In modern multiuse cutting tools with insert (e.g. with superfinishing edge or Wiper), such a cutting edge is made without documenting the basics for optimizing its dimensions. The article presents a generalized edge wear model in the surroundings of a rounded tip. In this sense the suggested solution is a novelty as it makes possible to specify the edge working conditions by modifying the edge corner and to adapt different tools for various cutting parameters established for creating a machined surface. The matter described in the article elaborates on patent number PL 173536 B1 where the usefulness of the solution for other tools (cutters, reamers, drills) was described. After the suggested works on the edge tip, due to the load along the cutting edge with the edge corner radius and along the cutting edge surface, always greater than zero and determining in both cases unit forces, it was possible to solve the mechanics of wear in the surroundings of the edge corner tip. The obtained model allows defining problems connected with the created unevenness on the machined surface. If plasticity phenomena in the cutting zone are taken into consideration, it can be proved that there is a group of wear marks after multiple passes of the edge tip in a disorganized random manner.

Keywords: Tool wear · Cutting edge · Corner radius · Turning

1 Introduction

The main objective of the article was to develop a new model of the course of wear of the clearance part located on corner top, which through its direct contact with the machined surface, exerts impact on creation of its surface.

As the cutting process progresses, marks of wear can be noticed on the tool's active surface [1, 2]. The marks image depends on how the tool angle is shaped geometrically. With a view on this, the most generic case of a tool with a circle-sector shaped edge was examined. The observed marks on a turning tool edge were depicted in Fig. 1. Three areas were marked on it, characterized by different mechanics and image of the marks. Differences between those images are seen between the areas I, II, III (Fig. 1). This depends on: cutting conditions, workpiece material of type, and strength of tool material. What needs to be mentioned here are the wear marks on the flank face or lack

© Springer Nature Switzerland AG 2020
G. M. Królczyk et al. (Eds.): IMM 2019, LNME, pp. 10–21, 2020.
https://doi.org/10.1007/978-3-030-49910-5_2

of a grooved mark on the tool face. In the cut layer, under the curvature of the cutting edge, there is a layer of plastic deformed material and it is up to approximately 0,2 mm depth. The thickness of the layer is marked in Fig. 1 with a dotted line, while the area is spotted.

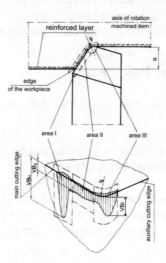

Fig. 1. Wear marks on the tool's working surfaces.

As the wear increases, a characteristic point where the cutting edge cuts the wrought layer appears in area I. The location of this point determines the cutting depth and a concentrated wear mark whose width is connected with the depth of the wrought layer appears around that point. The increased temperature in this area and the access of air from the edge open side intensify the wear growth, which manifests itself with a long local abrasion mark. In area II the abrasion adopts the shape of a rectangular strip that increases in width as the machining progresses. In area III the wear image is the result of highly complex mechanisms. In accordance with the previously discussed model, what is undergoing changes in the cutting zone is the layer thickness so that $h < h_{min}$. This happens in particular when the feed (f) is less than 0,2 mm per revolution, i.e. $f < 0,2$ mm/rev). [5, 6]. The mechanics of changes in tool edge apex geometry were examined for such geometric conditions.

A few articles have attempted to explain the cutting edge operation. The findings in [3] regarding the solution of turning tool wear are limited to value VB. This review covers cutting operation on its main cutting surface and reports on VBA, B, C, and N. Whereas, the auxiliary cutting surface describes VBr.

Value of VB wear does not influence the surface roughness parameters, while VBr directly modifies the unevenness set. The h adopted by the Author [4] does not take into consideration cutting force components, yet it considers only the geometric modeling. It specifically refers to cylindrical face milling, and not to turning [4].

The articles [7–9, 13–16] on cutting edge work and mathematical modeling of cutting forces describes the conditions during process milling. In the case of turning, the cutting edge has been described but without the wear tool being consumed and without affecting the rounding radius of the cutting edge r_n [10–12]. In the other publications [17–21] the values of tool wear are presented but without describing their actual impact on the mechanism of separation with the wear tool. The papers [22–26] presents the corners radius of the tool r_n, but without referring to the limited possibility of chip formation due to the minimum thickness of the cutting layer, as shown in the previously publications [5, 6] especially in the vicinity of the corner tip of the tool and its image.

2   Analytical Model of Tool Wear

The following initial conditions were adopted to create the analytical tool wear model:

1. The tool has an edge with radius r_ε,
2. Flank surface is circumscribed with angles α_p and α_f,
3. Rake surface is circumscribed with angle γ_f.

Apart from the geometric conditions it was assumed that, as a result of the wear, the tool is shortened by value KE. Angle λs would not contribute in any significant way to the discussion, but a correction including dislocation of the point on the tool edge apex has been included.

3   Geometry of the Tool Workpiece System

Generally, it can be assumed that the tool edge is a sector of a cylindrical surface with radius r_ε. This results from the fact that in the tool system, radius r_ε is shaped on a plane parallel to the basic plane. By grinding the main and auxiliary flank surface, the cylinder circumscribed on this edge is rotated in plane P_f by angle α_f and in plane P_p by angle α_p.

When turning, the workpiece is the cylinder surface with radius $R = D/2$. In relation to the cylinder, in accordance with the previous assumptions in the XYZ coordinate system, the cylinder circumscribed on edge with radius r_ε was spatially oriented. The coordinate system axes traverse in such a way so that they intersect the symmetry axes of the adopted cylinders (Fig. 2a).

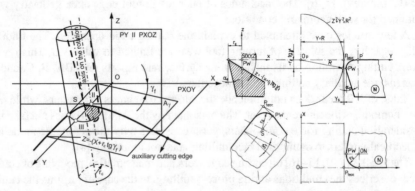

Fig. 2. a) simplified geometric model of cylinders intersecting, where one has radius R, b) location of the characteristic points on the tool edge in relation to the workpiece axis.

A circle, whose center lies in the place marked with S(Sx, Sy, Sz) point, was inscribed on the flank surface for $\gamma_f > 0$ (Fig. 2b). Axis of the cylinder circumscribed on the tool apex also passes through this point. The distance between the cylinders' axes, considering the wear values and impact of the rake angle, will be:

$$S_y = R + r_\varepsilon - \Delta \tag{1}$$

In this Eq. (1), apart from the sum of cylinders' radiuses, there is a value marked with Δ that is the sum of radial traverse, forced by the wear and Ys adjustment caused by the rake angle γ_f (Fig. 2b). This results from the fact that when the rake angle is $\gamma_f = 0$, point P_{max}, furthest on the edge tool apex, intersects with point P_{teor}, corresponding to the turning diameter. When rake angle $\gamma f \neq 0$ and tilt angle of the main cutting edge $\lambda s \neq 0$, relevant tool traverse, which decreases the distance between cylinders' axes by value Ys, needs to be taken into consideration. In the rear plane, in a section crossing through the apex point P_{max}, when the value of abrasion VBr on flank surface is known, radial tool wear KE can be calculated on basis of the following relation (2):

$$KE = \frac{VB'r \, \mathrm{tg}\, \alpha_p}{1 - \mathrm{tg}\, \gamma_p \, \mathrm{tg}\, \alpha_p} \tag{2}$$

And the equations system (3) of solution is:

$$\left. \begin{array}{c} Z^2 + Y^2 = R^2 \\ Z = -r_\varepsilon \, \mathrm{tg}\, \gamma_f \end{array} \right\} \tag{3}$$

We looked for point Y = Ys (Fig. 2), by which an analytical axis distance must be shortened. During wear tool VBr (Fig. 1) on the corner radius r_ε, which is expected value Ys, then shortened by on value KE (2).

Then we get the final Eq. (4) is therefore the sum of Eqs. (2) and (3).

From Eq. (4) Ys allows to explain the distance between the axis of the rotation and the substitute plane PY II PXOZ.

Whereas Y = R, when R is half workpiece diameter D (Fig. 2).

$$Y_s = R - \sqrt{R^2 - r_\varepsilon^2 \mathrm{tg}^2 \gamma_f} \tag{4}$$

Equation of the circle in a plane parallel to the basic plane P_r with center coordinates Sx = 0, Sy = R + r_ε − Δ, Sz = 0 in accordance with the system in Fig. 2a, can be written down as follows:

$$X^2 + [Y - (R + r_\varepsilon - \Delta)]^2 = r_\varepsilon^2 \tag{5}$$

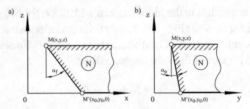

Fig. 3. Auxiliary diagrams for determining equations of the cylinder's generating lines: a) after rotating the line by angle α_f, b) after rotating by angle α_p.

On plane XOZ the cylinder is rotated by angle α_f, therefore the generating line of the cylinder should cross axis Z at this angle, as demonstrated in Fig. 3a. Similarly, in plane YOZ, the cylinder's generating line is rotated by angle α_p. Equation of the line passing through two points M and M′, which is also the equation of the cylinder generating line, can be drawn from the following relation (6) and (7):

$$\frac{X_0 - X}{0 - Z} = \operatorname{tg}\alpha_f \tag{6}$$

Where after conversion, expression for Z is determined

$$Z = \frac{X - X_0}{\operatorname{tg}\alpha_f} \tag{7}$$

Relevant dependences, but in YOZ plane, give (8)

$$\frac{Y_0 - Y}{0 - Z} = \operatorname{tg}\alpha_p \tag{8}$$

The final equation of line, crossing through points m and m′ and rotated by angle α_f and α_p can be written down as (9):

$$Z = \frac{X - X_0}{\operatorname{tg}\alpha_f} = \frac{Y_0 - Y}{\operatorname{tg}\alpha_p} \tag{9}$$

After determining from the system of equations

$$\left.\begin{array}{l} X = X_0 + Z\operatorname{tg}\alpha_f \\ Y = Y_0 + Z\operatorname{tg}\alpha_p \end{array}\right\} \tag{10}$$

And after replacing in the circle Eq. (3) the following is obtained Eq. (11)

$$\left(X_0 + Z\operatorname{tg}\alpha_f\right)^2 + \left(Y_0 + Z\operatorname{tg}\alpha_p - R - r_\varepsilon + KE + Ys\right)^2 = r_\varepsilon^2 \tag{11}$$

When each point on the generating line also lies on the cylinder's surface in the adopted tool - workpiece system (Fig. 2a), and the equation of the cylinder circumscribing the workpiece machined surface has the following form:

$$\left.\begin{array}{l} Z^2 + Y^2 = R^2 \\ \left(X + Z\cdot tg\alpha_f\right)^2 + \left(Y + Z\cdot tg\alpha_p - R - r_\varepsilon + KE + Ys\right) = r_\varepsilon^2 \end{array}\right\} \tag{12}$$

Then the solution is a closed space curve which is the edge of intersection of intersecting cylinders. The equations obtained as a result of equation conversion (2) expression

$$Y = \sqrt{R^2 - Z^2} \tag{13}$$

replaced in Eq. (12) cannot be ousted and for this reason the system of Eqs. (12) is difficult to solve. The solution can be facilitated by adopting known values of variable Z from the range of $-R < Z < +R$ Z values, for which discrimination of quadratic equation is greater than zero, in the field of X, Y, Z data for the cylinders intersecting curve.

4   Simplified Case of Tool - Workpiece System

Due to the complex form of general Eqs. (12), a different geometric model of the tool - workpiece system was adopted (Fig. 2a). When the ratio of radiuses r_ε to R is low, then it can be assumed that the cylinder circumscribed on the apex radius is cut with PY plane, parallel to PXOZ place. The PY plane was placed at a distance $Y = R$ from the beginning of the coordinates system. Just like previously, a cylindrical surface rotated at angles α_f and α_p can be led through the edge tool, before the wear out of the object traversed by the distance $Y = R$ to the workpiece (Fig. 2a). Therefore, the cylindrical surface equation will have the following form (14)

$$\left(X + Z\,tg\,\alpha_f\right)^2 + \left(Y + Z\,tg\,\alpha_p - R - r_\varepsilon + KE + Ys\right)^2 = r_\varepsilon^2 \tag{14}$$

where all variables are the same, just as in Eq. (12).

After replacement of value $Y = R$, i.e. after intersecting the cylindrical surface (14) with place PY, the following is obtained

$$\left(X + Z\,tg\,\alpha_f\right)^2 + \left(Z\,tg\,\alpha_p - r_\varepsilon + KE + Ys\right)^2 = r_\varepsilon^2 \tag{15}$$

After multiplying the components and comparing the general equation of a 2nd order curve

$$AX^2 + BXZ + CZ^2 + DX + EZ + F = 0 \tag{16}$$

With development of Eq. (14) the following was obtained:

$$A = 1 \tag{17}$$

$$B = 2\,tg\,\alpha_f \tag{18}$$

$$C = tg^2\alpha_p + tg^2\alpha_f \tag{19}$$

$$D = 0 \tag{20}$$

$$E = tg\,\alpha_p(KE + Ys - r_\varepsilon) \tag{21}$$

$$F = (KE + Ys - r_\varepsilon)^2 - r_\varepsilon^2 \tag{22}$$

then the examined equation invariants (16)

$$w = \begin{vmatrix} A & B \\ B & C \end{vmatrix} > 0 \tag{23}$$

$$W = \begin{vmatrix} A & B & D \\ B & C & E \\ D & E & F \end{vmatrix} < 0 \tag{24}$$

therefore, the value as well

$$AW < 0 \tag{25}$$

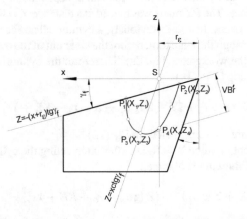

Fig. 4. Auxiliary diagram for calculating coordinates of the ellipse's characteristic points.

As a result, the obtained curve is the equation of the actual ellipse. After calculating the necessary auxiliary data and inserting them into the ellipse Eq. (15) characteristic points P_1, P_2, P_3 and P_4 can be established that describe the ellipse section, limited with rake surface, as illustrated in Fig. 4.

When the rake angle γ_f is greater than zero, equation of line which will be the edge of intersection of PY place with the plane crossing through the rake surface needs to be inserted into (26), i.e.

$$Z = -(X + r_\varepsilon)\,\mathrm{tg}\,\gamma_f \tag{26}$$

Then the following equation is obtained

$$X^2\left[\left(1 - r_\varepsilon\,\mathrm{tg}\,\alpha_f\,\mathrm{tg}\,\gamma_f\right)^2 + \mathrm{tg}^2\alpha_p\,\mathrm{tg}^2\,\gamma_f\right] + X\left[-\left(2r_\varepsilon\left(1 - \mathrm{tg}\,\alpha_f\,\mathrm{tg}\,\gamma_f\right)\mathrm{tg}\,\alpha_f\,\mathrm{tg}\,\gamma_f\right.\right.$$
$$+ 2\,\mathrm{tg}\,\alpha_p\,\mathrm{tg}\,\gamma_f\left(-r_\varepsilon\,\mathrm{tg}\,\alpha_f\,\mathrm{tg}\,\gamma_f - r_\varepsilon + KE + Ys\right) + r_\varepsilon^2\,\mathrm{tg}^2\,\alpha_f\,\mathrm{tg}^2\,\gamma_f$$
$$+ \left(-r_\varepsilon\,\mathrm{tg}\,\alpha_f\,\mathrm{tg}\,\gamma_f - r_\varepsilon + KE + Ys\right)^2 - r_\varepsilon^2 = 0 \tag{27}$$

After introducing auxiliary designations

$$A = \left[1 - r_\varepsilon \, \text{tg} \, \alpha_f \, \text{tg} \, \gamma_f\right]^2 + \text{tg}^2 \, \alpha_p \, \text{tg}^2 \, \gamma_f \tag{28}$$

$$B = \left[-2r_\varepsilon\left(1 - \text{tg} \, \alpha_f \, \text{tg} \, \gamma_f\right)\text{tg} \, \alpha_f \, \text{tg} \, \gamma_f + 2 \, \text{tg} \, \alpha_p \, \text{tg} \, \gamma_f\left(-r_\varepsilon \, \text{tg} \, \alpha_f \, \text{tg} \, \gamma_f - r_\varepsilon + KE + Ys\right)\right] \tag{29}$$

$$C = r_\varepsilon^2 \, \text{tg}^2 \, \alpha_f \, \text{tg}^2 \, \gamma_f + \left(-r_\varepsilon \, \text{tg} \, \alpha_f \, \text{tg} \, \gamma_f - r_\varepsilon + KE + Ys\right)^2 - r_\varepsilon^2 \tag{30}$$

Equation (27) is reduced to the form

$$A X^2 + BX + C = 0 \tag{31}$$

Task roots (31)

$$X_{1,2} = \frac{-B \mp \sqrt{B^2 - 4AC}}{2A} \tag{32}$$

introduced into expression (26), makes it possible to calculate coordinates Z1 and Z2 of points P_1 and P_2 from Fig. 4. P_4 point coordinates are determined after introducing into Eq. (15) cordinate $X_4 = 0$ for point P_4 and a second order equation is obtained:

$$Z_4^2\left(\text{tg}^2 \, \alpha_f + \text{tg}^2 \, \alpha_p\right) - 2 Z_4 \, \text{tg} \, \alpha_p(r_\varepsilon - KE - Ys) + (r_\varepsilon - KE - Ys)^2 - r_\varepsilon^2 = 0 \tag{33}$$

For point P_3, corresponding to intersection of the ellipse with its longer axis, after substituting equation of the line in Eq. (15)

$$Z = -X \, \text{ctg} \, \alpha_f \tag{34}$$

Overlapping the axis, the following equation is obtained

$$X^2 \text{tg}^2 \alpha_p \, \text{ctg}^2 \, \alpha_f + 2X \, \text{tg} \, \alpha_p \, \text{ctg} \, \alpha_f(r_\varepsilon - KE - Ys) + (r_\varepsilon - KE - Ys)^2 - r_\varepsilon^2 = 0 \tag{35}$$

Roots of Eqs. (33) and (35) are the sought coordinates of points P_3 and P_4 of the ellipse fragment (Fig. 4). When angle α_f is low, then the difference between the coordinates of points P_3 and P_4 is negligible in the calculations.

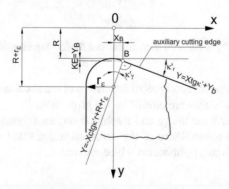

Fig. 5. Auxiliary diagram for calculating width of the abrasion mark on the transitory flank surface.

The auxiliary cutting edge is rectilinear and transforms into a curved edge in point B, as demonstrated in Fig. 5. When the axial tool wear KE obtains value higher than Y_B, it means that when

$$KE \geq Y_B \text{ dla } Y_B = r_\varepsilon \left(1 - \cos \kappa_r'\right) \tag{36}$$

then after substituting Eq. (2) in (36) the following is obtained

$$\frac{VBr \, \text{tg} \, \alpha_p}{1 - \text{tg} \, \gamma_p \, \text{tg} \, \alpha_p} \geq r_\varepsilon \left(1 - \cos \kappa_r'\right) \tag{37}$$

In this expression VBr is the height of abrasion on the auxiliary flank surface. The obtained wear mark width X1 + X2 (Fig. 4) will then be greater than results from solution (32), which is due to the fact that as the wear progresses, the point located on the circle moves onto the auxiliary edge.

5 Experimental Verification of Model Correctness

The interaction of friction surfaces, the cutting edge and machined material are accompanied by mutual wear. The abrasion on the corner radius is concentrated in a specific area. However, the phenomenon of friction in the area of common contact results in an unevenness image, as presented in Fig. 6. Growing abrasion marks are visible in a 50-times magnification.

This does not apply to the cracks depth, which are at least a grade smaller. This is irrelevant for our considerations as it would only hinder illustrating the issue with the descriptive method.

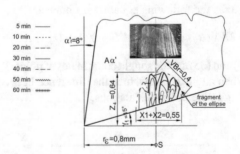

Fig. 6. Wear lands on transitional tool flank. Magnification 50×

In this work, the experimentally recorded curves of the abrasion ellipses corresponding to subsequent wear stages are plotted on the projection.

In Fig. 7 the corner wear image and machined surface fragment were put together. This figure shows the resemblance of the observed images that were collected while illuminating them with two light beams – blue and red.

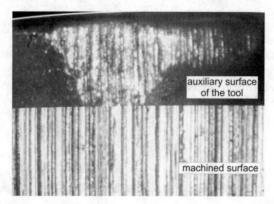

Fig. 7. Composition of two images: image of wear traces on the tip of cutting part and corresponding image of traces on the machined surface

It should be noted that many more details are made visible with the optic method, in contrast to 2D or 3D records on profilometers. The wear marks measured by increasing cutting time present a growth in the abrasion dimensions, which was presented in Fig. 8 as an example.

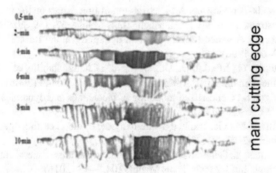

Fig. 8. Summary of wear marks, along the rounding of the corner radius for different cutting times

6 Conclusions

The vast amount of literature data concerning VBr changes that are dependent on time has encouraged the Authors to write down the following conclusions, focusing especially on some new aspects:

1. Solutions of the equations allow us to generalize the model of contact of two warped rollers and changes in the location of their axes towards one another due to the progressing tool wear.

2. The distance between the axes remains unchanged and wear of one of them changes the contact area.
3. As seen on the model, unevennesses in the diffusion area that make up the surface roughness have been modified.
4. This results from multiple passes of the machined surface across the abrasion width at the edge of top (Fig. 7 and 8).

References

1. Brammertz, P.H.: Die Enstehung der Oberflächenerauheit beim Feindrehen, Industrie Anzeiger, 2 (1961)
2. Gieszen, C.A.: Bas seitliche Verguetschen des Werkstoffes auf den Oberfläehen feingedrehter Werkstücke, Fertigung, 5 (1971)
3. Mikołajczyk, T., Nowicki, K., Kłodowski, A., Pimenov, D.Y.: Neural network approach for automatic image analysis of cutting edge wear. Mech. Syst. Signal Process. **88**, 100–110 (2017)
4. Pimenov, D.Y.: Geometric model of height of microroughness on machined surface taking into account wear of face mill teeth. J. Frict. Wear **34**(4), 290–293 (2013)
5. Storch, B., Zawada-Tomkiewicz, A.: Distribution of unit forces on the tool edge rounding in the case of finishing turning. Int. J. Adv. Manuf. Technol. **60**(5–8), 453–461 (2012)
6. Storch, B., Zawada-Tomkiewicz, A.: Distribution of unit forces on the tool nose rounding in the case of constrained turning. Int. J. Mach. Tools Manuf. **57**, 1–9 (2012)
7. Altintas, Y., Lee, P.: A general mechanics and dynamics model for helical end mills. CIRP Ann. **45**(1), 59–64 (1996)
8. Altintas, Y., Stepan, G., Merdol, D., Dombovari, Z.: Chatter stability of milling in frequency and discrete time domain. CIRP J. Manuf. Sci. Technol. **1**, 35–44 (2008)
9. Altintas, Y., Engin, S.: Generalized modeling of mechanics and dynamics of milling cutters. CIRP Ann. **50**(1), 25–30 (2001)
10. Altintas, Y., Kilic, Z.M.: Generalized dynamic model of metal cutting operations. CIRP Ann. – Manuf. Technol. **62**, 47–50 (2013)
11. Kilic, Z.M., Altintas, Y.: Generalized mechanics and dynamics of metal cutting operations for unified simulations. Int. J. Mach. Tools Manuf. **104**, 1–13 (2016)
12. Kilic, Z.M., Altintas, Y.: Generalized modeling of cutting tool geometries for unified process simulation. Int. J. Mach. Tools Manuf. **104**, 14–25 (2016)
13. Lee, P., Altintas, Y.: Prediction of ball-end milling forces from orthogonal cutting data. Int. J. Mach. Tools Manuf. **36**(9), 1059–1072 (1996)
14. Larue, A., Altintas, Y.: Simulation of flank milling processes. Int. J. Mach. Tools Manuf. **45**, 549–559 (2005)
15. Kaymakci, M., Kilic, Z.M., Altintas, Y.: Unified cutting force model for turning, boring, drilling and milling operations. Int. J. Mach. Tools Manuf. **54–55**, 34–45 (2012)
16. Budak, E., Ozlu, E.: Development of a thermomechanical cutting process model for machining process simulations. CIRP Ann. – Manuf. Technol. **57**, 97–100 (2008)
17. Altintas, Y., Jin, X.: Mechanics of micro-milling with round edge tools. CIRP Ann. – Manuf. Technol. **60**, 77–80 (2011)
18. Liu, Q., Altintas, Y.: On-line monitoring of flank wear in turning with multilayered feed-forward neural network. Int. J. Mach. Tools Manuf. **39**, 1945–1959 (1999)
19. Berenjia, K.R., Karaa, M.E., Budak, E.: Investigating high productivity conditions for turn-milling in comparison to conventional turning. Procedia CIRP **77**, 259–262 (2018)

20. Bagherzadeh, A., Budak, E.: Investigation of machinability in turning of difficult-to-cut materials using a new cryogenic cooling approach. Tribol. Int. **119**, 510–520 (2018)
21. Olgun, U., Budak, E.: Machining of difficult-to-cut-alloys using rotary turning tools. Procedia CIRP **8**, 81–87 (2013)
22. Jin, X., Altintas, Y.: Prediction of micro-milling forces with finite element method. J. Mater. Process. Technol. **212**, 542–552 (2012)
23. Jin, X., Altintas, Y.: Slip-line field model of micro-cutting process with round tool edge effect. J. Mater. Process. Technol. **211**, 339–355 (2011)
24. Karaguzela, U., Uysalb, E., Budak, E., Bakkal, M.: Effects of tool axis offset in turn-milling process. J. Mater. Process. Technol. **231**, 239–247 (2016)
25. Celebia, C., Özlüb, E., Budak, E.: Modeling and experimental investigation of edge hone and flank contact effects in metal cutting. Procedia CIRP **8**, 194–199 (2013)
26. Budak, E., Ozlu, E., Bakioglu, H., Barzegar, Z.: Thermo-mechanical modeling of the third deformation zone in machining for prediction of cutting forces. CIRP Ann. – Manuf. Technol. **65**, 121–124 (2016)

Evaluation of Deflection of Thin-Walled Profile During Milling of Hardened Steel

Jakub Czyżycki[✉] and Paweł Twardowski

Institute of Mechanical Technology, Poznan University of Technology, Ul. Piotrowo 3,
60-965 Poznań, Poland
jakub.r.czyzycki@doctorate.put.poznan.pl

Abstract. The aim of the research was to evaluate deflection of thin-walled profile made of hardened steel during milling with variable cutting parameters. During the cutting of thin-walled elements, there is a probability of their deflection, which is connected with dimensional errors of the made elements. The tests included measurement of the components of the total cutting force using a piezoelectric accelerator, and deflections of a thin-walled element using a laser position sensor during milling of thin-walled elements at different cutting speeds and feeds in hardened steel using end flat mill. The tests showed the minimum wall thickness for the tested material and the influence of cutting parameters on their deflection.

Keywords: Milling · Thin-walled elements · Hardened steel · Deflection

1 Introduction

We consider thin-walled elements to be objects whose height-to-thickness ratio is large. The author [1] specifies a 30: 1 height to thickness ratio as large. Examples of the use of thin-walled elements in machine construction are impellers and monolithic ribs [2].

During milling of thin-walled elements, the cutting force occurring deflects elastically and plastically the workpiece, hence the need to take deflection into account when using this type of machining. Under the influence of elastic deformation, the workpiece is deformed, which leads to changes into the thickness of machined layer and thus dimensional accuracy of finished components decreased. In addition, the resulting vibrations deteriorate the quality of the milled surface. On the other hand, plastic deformations may leave unwanted own stresses in the material which will require removal in the annealing processes.

Problem of deformation of thin-walled elements is common in industry and cause many problems. As an example is difficulty or the impossibility of assembly construction elements during their production [1, 3].

G. M. Królczyk et al. (Eds.): IMM 2019, LNME, pp. 22–32, 2020.
https://doi.org/10.1007/978-3-030-49910-5_3

According to many sources, the improvement of the quality of thin-walled elements has been improved by many ways [1, 3–5]:

- optimization of machining strategy,
- appropriate selection of feed per tooth f_z and cutting width a_e in order to minimize the component of the cutting force which is perpendicular to the thin-walled wall of the workpiece,
- use of higher cutting speeds – HSM (High Speed Machining), HSC (High Speed Cutting),
- bear in mind the possible stresses and deformations that occur when clamping the workpiece on the machine tool.

The basic procedure that allows to avoid deformation of the walls while selecting machining strategies and selecting cutting parameters is the use of a low coefficient of depth of cut to a_p/a_e width and high v_c cutting speed, which allows shortening the time of tool contact with the workpiece, while the number of transitions is already selected based on the dimensions of the element [1]. In addition to these studies [1, 6, 7] the positive effect of up milling at the walls without rigid support has been proved.

The HSM has been acknowledged well known and widely used technique for the production of thin-walled items used for aerospace applications. Increasing the cutting speed v_c allows to reduce the cutting forces. It allows to make a smaller number of integral parts instead of many elements that were combined together. It requires machine tools with electro-spindle or pneumatically driven spindle [1, 6].

In order to eliminate the deflection associated with fixing the workpiece, the author [8] recommends using vacuum fixtures or self-adjusting fixtures, while the disadvantage is high price and lack of universality.

There are other techniques to compensate for the deflection of thin-walled elements, such as measuring the cutting forces in real time and deflecting the direction of the tool depending on this force. However, this method has not yet been applied in industry [9].

The next important stage of prediction of deflection of thin-walled objects under the influence of cutting forces includes model tests on finite elements to measure the deflection of the object at various stages of the tool path. In study [9] authors were able to reduce deformation of thin wall elements, vibrations and cutting forces by using multilayer active compensation. The difference in element deformation between the simulation and the experiment in this case was only 0.0062 mm, which proves the usefulness of this method. Studies have proven that it is very important to take into account the cutting forces and clamping forces of thin-walled elements during their machining.

2 Experimental Details

Tool steel 55NiCrMoV6 (WNL) with hardness 54 HRC, whose chemical composition is shown in Table 1, was tested.

Table 1. Chemical composition of the sample made of 55NiCrMoV6 steel.

C	Mn	Si	P	S	Cr	Ni	Mo	W	V	Co	Cu
0.5–0.6	0.5–0.8	0.15–0.4	Max. 0.03	Max. 0.03	0.5–0.8	1.4–1.8	0.15–0.25	Max. 0.03	Max. 0.1	Max. 0.03	Max. 0.03

The machining was carried out with a monolithic milling cutter as can be seen in Fig. 1 [10]. The milling cutter was made of fine-grained cemented carbide of dimensions: diameter of $D = 10$ mm and number of blades $z = 4$. The main cutting edge angle $\lambda s = 40°$ and rake angle $\gamma 0 = 5°$ respectively.

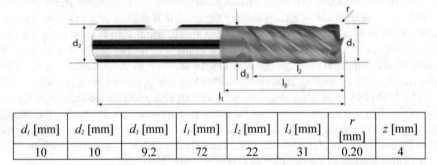

d_l [mm]	d_2 [mm]	d_3 [mm]	l_l [mm]	l_2 [mm]	l_3 [mm]	r [mm]	z [mm]
10	10	9.2	72	22	31	0.20	4

Fig. 1. Flat end mill: FRAISA P9900 450 [10].

In the rectangular block, 6 walls were made, 2 mm thick, 8 mm high, and 20 mm long. The walls were set parallel to each other. Figure 2 shows the walls made. The sample was mounted on a dynamometer which was mounted on the machine table.

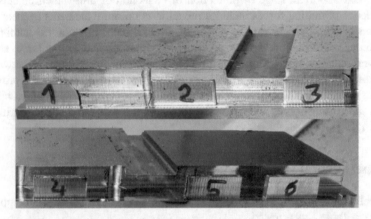

Fig. 2. Sample of WNL steel with prepared thin walled elements.

The experimentation was carried out on a DACKEL MAHO three-axis machining center model DMC 70 V. Figure 3 shows the machine tool along with the measuring track for forces and deflections.

Fig. 3. DMC 70 V machine tool from DECKEL MAHO along with a measuring track..

Prepared samples were divided into two groups. The first group was milled with a constant feedrate at a variable cutting speed value. The second group had a constant value of cutting speed during milling and variable feed. The process itself was designed to reduce the thickness of each wall from 2 mm to 0.5 mm with the cutting width $a_e = 0.15$ mm and $a_p = 8$ mm, so 10 passes were made on each wall. In both cases, down milling was used without the use of a cooling liquid. Table 2 shows the cutting parameters used in the test.

Table 2. Cutting parameters used in the tests.

Wall number	Cutting speed v_c [m/min]	Feed per tooth f_z [mm]	Rotation speed n [rev/min]	Cutting feedrate v_f [mm/min]
1.	50	0.05	1592	318
2.	200	0.05	6370	1274
3.	300	0.05	9550	1910
4.	100	0.02	3183	255
5.	100	0.05	3183	637
6.	100	0.1	3183	1273

The deflections measurements of the tested wall were made with the Micro-Epsilon optoNCDT ILD1700-10 LL sensor (Fig. 4), measuring deflections in the direction perpendicular to the tool feed. Sensor with a measuring range of 10 mm and having least count of 0,5 μm was used.

To analyze the obtained results, the Analyzer program was used in which, the deflection of the wall was determined manually for each result. Deflection of the wall was

Fig. 4. Micro-Epsilon optoNCDT ILD1700-10 LL deflection sensor.

determined as the difference from the zero position of the sensor to the maximum wall deflection value measured on the basis of the mean value of the three highest elevations in the graph as can be seen in Fig. 5.

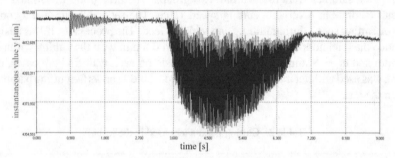

Fig. 5. Deflection of the first wall during the tenth transition of the end mill.

The measurement of the components of total cutting force during milling was carried out using a three-component dynamometer platform connected to charge amplifiers, an analog-digital converter and a PC equipped with a data processing and analysis program. Figure 6 shows the diagram of the deflection measurement track and the components of the cutting forces, the coordinate system refers to the measuring directions of the measuring platform forces.

Each time after making 5 passes and 10 passes, the wall thickness was measured with a measuring probe made by Renishaw (Fig. 7).

3   Results and Discussion

Figures 8 and 9 show graphs for measuring the thickness of six walls, of which three were machined at constant feed and variable cutting speed, and three were machined with variable feed and constant cutting speed. The results show wall thickness after 5 milling passes, 0.75 mm of the material was machined, and after 10 passes 1.5 mm was

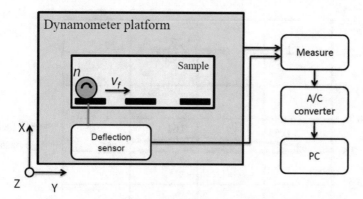

Fig. 6. Diagram of the measuring track.

Fig. 7. Measuring probe.

machined, where the initial value of the wall thickness was 2 mm. As can be seen from the first 5 passes, this thickness is very similar to the one assumed, while after 10 passes where the thickness of the wall in relation to its height was low, these values are different from nominal. All walls are thicker than they should be after milling, the second and fifth walls have as much as 0.61 mm thick.

This testifies to the creation of elastic deformation during cutting, in particular in passages 5–10, which caused a reduction in the thickness of the cutting layer in each of the transitions and gave a total smaller value of the final thickness. The red lines in the graph refer to nominal values after 5 passes (1.25 mm) and 10 passes (0.5 mm).

As can be seen in both cases, increasing the cutting speed or feed rate does not affect the wall thickness unambiguously. Initially, this value increases with the increase of cutting parameters, while with further increase of the cutting speed or feed, the wall thickness decreases, returning to the initial value.

Deflection of each wall was measured by displacement sensor. A list of results for walls 1–3 is shown in Fig. 10. For each wall, you can see a rapid increase in the deflection after the fifth pass. This allows us to determine the limit wall thickness for this material and cutting parameters below which its deflection is significant and as shown in Figs. 8 and 9, it affects the resulting wall thickness.

The first milled wall with the lowest cutting speed ($v_c = 50$ m/min) obtained the smallest deflection values over its entire run compared to the next two, the characteristics of which overlap the most. From this it can be concluded that in a certain range of cutting

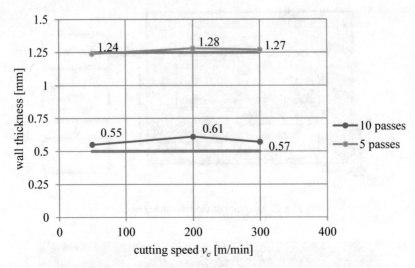

Fig. 8. List of wall thicknesses 1–3 milled with constant feed f_z and variable cutting speed v_c.

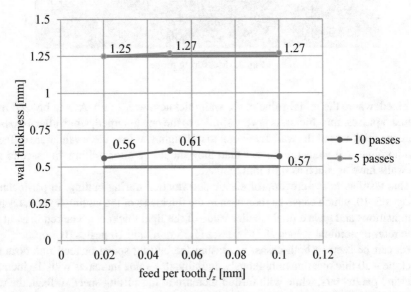

Fig. 9. List of wall thicknesses 4–6 milled with constant cutting speed v_c and variable feed f_z.

speeds wall deflection during milling will be constant, and in order to reduce it, we must reduce the cutting speed, while the effect of its increase above 300 m/min should be additionally checked.

The results of displacements of walls 4–6 are shown in Fig. 11, in contrast to walls 1–3. In this case, significant anomalies can be noticed which make it difficult to analyze the results obtained and the reason for their occurrence, probably due to problems with the measurement path or its operators. However, the difference in the deflection of the

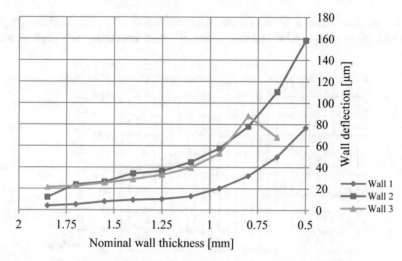

Fig. 10. List of displacement results of walls 1–3 during milling.

wall number 6 as compared to the walls 4 and 5 can be noticed, to which the variable feed did not affect virtually at all. That also mean that there is a certain range of feed per tooth that will obtain constant wall deflection.

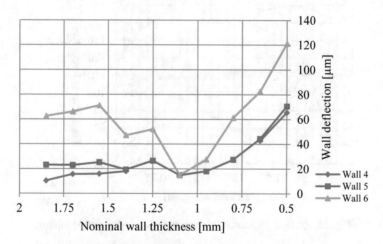

Fig. 11. List of displacement results of walls 4–6 during milling.

The next parameter which has been examined is the value of the feed normal force *FfN* during cutting. It is a force acting perpendicular to the surface of the wall and the direction of feed in down milling, so among all the components of the cutting force, it should have the greatest impact on its deflection. Figures 12 and 13 show a comparison of normal thrust force with respect to the nominal wall thickness. Normal feed force values were determined from the results that were processed in the Analyzer program as the root mean square (RMS) on the section on which the cutting was performed.

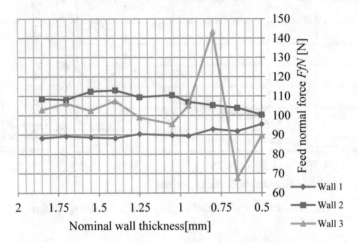

Fig. 12. Normal feed force *FfN* values for walls 1–3 during milling.

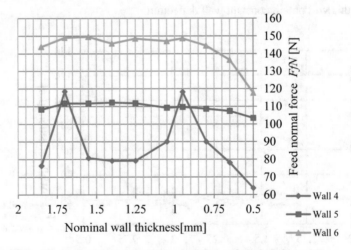

Fig. 13. Normal feed force *FfN* values for walls 4–6 during milling.

Analyzing the characteristics of wall deflections in Figs. 10 and 11 we would expect an decrease in the value of the feed normal force above 6–7 pass of every wall (less than 1 mm thick), and that can be observed on diagrams from Figs. 12 and 13. As the wall thickness decreases, the value of the normal thrust force decreases. It is evident that below 1 mm of thin-walled element made of hardened steel used in the study, regardless of the during parameters (v_c or f_z), the value of their deflection increases significantly. A decrease in feed normal force is a sign of a decrease in the workpiece rigidity and the formation of elastic deformations.

4 Conclusions

The following conclusions can be made on the basis of the tests and results:

- For the parameters specified in the test the thickness of 1 mm is the limit thickness below which the value of deflection increases significantly, which is reflected in the lack of cutting the given layer of material,
- From the cutting parameters that were changed, i.e. is the cutting speed v_c and the feed per tooth f_z the former affected the deflection of the wall to a greater extent, while the final value of wall thickness after cutting in both cases, despite the too large thickness, changed in the same way with increase in the main parameter.
- Similar deflection values for walls 2 and 3, and 4 and 5 indicate a certain range of cutting parameters (cutting speed v_c and feed per tooth f_z) where the deflection value is constant. This allows optimizing machining by gaining machining time by examining these ranges for a given application.
- A relationship has been highlighted in which below the wall thickness of 1 mm the value of the feed normal force decreases. This is due to the material being lost in stiffness and increasing its susceptibility to deflection. Below 1 mm of wall thickness, elastic deformations began to occur, which are visible both in the decrease in cutting force and the lack of obtaining nominal wall dimensions. Despite the use of hardened steel with a hardness of 54HRC, the limit value of wall thickness that allows maintaining dimensional accuracy does not differ from the aluminum alloy used in the work [3] under similar test conditions.
- In order to better illustrate the effect of cutting parameters on the thickness of the obtained wall after milling, it would be necessary to make several additional elements with a greater range of cutting speed and feeds, out of three walls for each parameter, the results are not conclusive.

Referencing

1. Zębala, W.: Errors minimalisation of thin-walled parts machining. Mach. Eng. **15**(3), 45–54 (2010)
2. Huang, N., Bi, Q., Wang, Y., Sun, Ch.: 5-axis adaptive flank milling of flexible thin-walled parts based on the on-machine measurement. Int. J. Mach. Tools Manuf **84**, 1–8 (2014)
3. Kuczmaszewski, J., Pieśko, P., Zawada-Michałowska, M.: Analysis of cutting speed influence on the deformation of thin-walled elements made of aluminium alloy EN AW-2024 after milling. Mechanik Nr. **8–9**, 1066–1067 (2016)
4. https://www.sandvik.coromant.com/pl-pl/knowledge/milling/Pages/default.aspx
5. Włodarczyk, M.: Analiza wpływu sił skrawania oraz zamocowania na poziom naprężeń w aspekcie grubości ścianek wybranej konstrukcji kieszeniowej. Postęp Nauki i Techniki Nr. **8**, 82–91 (2011)
6. Bałoń, P., Rejman, E., Smusz, R., Kiełbasa, B.: High speed machining of the thin-walled aircraft constructions. Mechanik Nr. **8–9**, 726–729 (2017)
7. Masmali, M,. Mathew, P.: An analytical approach for machining thin-walled worpieces. In: 16th CIRP Conference on Modelling of Machining Operations, CIRP, vol. 58, pp. 187–192 (2017)

8. Campa, F.J., Lopez, L.N., de LAcalle, L., Celaya, A.: Chatter avoidance in the milling of thin floors with bull-nose end mills: model and stability diagrams. Int. J. Mach. Tools Manuf. **51**, 43–53 (2011)

9. Chen, W., Xue, J., Tang, D., Ch, H., Qu, S.: Deformation prediction and error compensation in multilayer milling processes for thin-walled parts. Int. J. Mach. Tools Manuf **49**, 859–864 (2009)

10. http://webshop.fraisa.ch/?lng=en

The Surface Texture of Ti6Al4V Titanium Alloy Under Wet and Dry Finish Turning Conditions

Kamil Leksycki[✉] and Eugene Feldshtein

Faculty of Mechanical Engineering, University of Zielona Gora, 4 Prof. Z. Szafrana Street,
65-516 Zielona Gora, Poland
k.leksycki@ibem.uz.zgora.pl

Abstract. The paper presents the results of investigations of the surface texture of Ti6Al4V titanium alloy after finish turning in dry and wet cooling conditions. Variable cutting speeds and feed rates with the constant depth of cutting were used. The research plan was realized with the use of the PSI (Parameter Space Investigation) method, which allows carrying out research tests with minimization of the number of experimental points. The measurements were made with the use of Keyence VHX-6000 digital microscope. The aim of the study was to determine the influence of cutting parameters on the surface texture of Ti6Al4V titanium alloy when finish turning under dry and wet cooling conditions. The feed range studied was 0.05–0.4 mm/rev, cutting speed was 40–120 m/min and the stable cutting depth of 0.5 mm was used. Under dry turning, both the Ra value of the surface roughness and the intensity of the cutting speed and feed rate influence are lower in comparison with wet turning. Compared to wet machining, a higher number of surface defects caused by adhesive bonds were observed under dry turning. Favorable values of roughness parameters were obtained using middle cutting speeds and feed rates of the range tested.

Keywords: Surface texture · Finish turning · Wet and dry cutting · Ti6Al4V titanium alloy

1 Introduction

The materials most commonly used in medicine are: stainless steels, nickel-based alloys, titanium alloys, cobalt-chromium alloys and others [1]. Among these groups, 316L stainless steel and Ti-Al-V, Co-Cr-Mo and Ni-Ti alloys are the most frequently used and studied ones [2, 3].

Ti6Al4V alloy is the most commonly used titanium alloy, which is successfully used for dental implants, bone fixing devices, artificial heart valves or elements in orthodontic surgery [4–6]. Ti6Al4V titanium alloy has favorable mechanical properties; it is chemically inert in the human body environment and has very good corrosion resistance, making it a biocompatible material [7–11]. Ti6Al4V titanium alloy is characterized by low thermal conductivity, low modulus of elasticity and high chemical reactivity. The material properties of Ti6Al4V titanium alloy result in a high level of vibration during machining. Therefore, low cutting speeds are recommended for machining [12–15].

© Springer Nature Switzerland AG 2020
G. M. Królczyk et al. (Eds.): IMM 2019, LNME, pp. 33–44, 2020.
https://doi.org/10.1007/978-3-030-49910-5_4

High cutting speeds are desirable in order to increase productivity. However, they generate large amounts of heat and high temperatures in the cutting zone, leading to faster tool wear and reducing surface quality. In the case of titanium and its alloys, this problem is more serious because of the low thermal conductivity. About 80% of the heat generated in the cutting zone flows directly to the tool [16]. A worn tool causes the deterioration of surface roughness, which is always treated as an indicator of the quality of a semi-finished or finished product. Although Ti6Al4V titanium alloy is a commonly known material, it belongs to the group of difficult to be machined and has a complicated chip formation process [17–19]. Chip shapes during the machining of titanium alloys are directly related to the cutting parameters, and the greatest differences can be observed when machining with higher feed rates and higher cutting speeds [15].

The integrity of the surface machined is assessed in terms of surface topography and defects [20, 21], as well as residual stresses [22]. Sartori et al. [22], turned Ti6Al4V titanium alloy with a depth of $a_p = 0.25$ mm under dry and cryogenic cooling conditions. The machining was carried out with a variable feed rate and a constant depth and speed of cutting. The authors noted that in comparison with dry turning, cryogenic machining ensured a smaller number of defects on the machined surface and higher values of residual stresses, and a deterioration of surface topography was also observed.

Bordin et al. [23] compared dry and cryogenic cooling in the turning process with a depth of $a_p = 0.25$ mm of Ti6Al4V titanium alloy, where they applied variable cutting speeds and feed rates as well as a constant cutting depth. The results showed that cryogenic cooling provides better performance compared to dry and wet machining by reducing tool wear, improving surface quality and increasing chip breakability. Under dry machining conditions, more surface defects and severe tool wear occurred.

The growing method of cutting is cryogenic cooling, which increases the machining efficiency and is safe for the environment. Mia et al. [24] investigated effect of multi-objective optimization and life cycle changes when eco-friendly cryogenic N_2 assisted turning of Ti-6Al-4V. Jamil et al. [25] researched effects of hybrid Al_2O_3-CNT nanofluids and cryogenic cooling on machining of Ti–6Al–4V. Mia and Dhar [26] studied influence of single and dual cryogenic jets on machinability characteristics when turning of Ti-6Al-4V.

Liang et al. [27] examined the surface topography and its deterioration as the result of tool wear during the turning of Ti6Al4V titanium alloy under dry conditions with constant cutting speeds (75 m/min), feed (0.1 mm/rev) and depth of cutting (0.5 mm). Unwanted surface defects such as feed marks, plowing grooves, side flow, plastic flow, adhered material particles, surface tearing, surface burning, scratch marks were determined basing on of tool wear features and resulting thermal and mechanical loads. The relationship between the evolution of surface topography and the effects of tool wear was shown.

Liang et al. [28] studied the behavior and mechanical properties of Ti6Al4V titanium alloy during turning under high pressure cooling conditions. The results were compared to dry machining. The surface roughness parameters (arithmetical mean height Sa and developed interfacial area ratio Sdr) decreased under high-pressure cooling conditions. The depth of the plastic layer of the material decreased with increasing pressure of the cooling medium. Also, the content of β-phase of Ti6Al4V titanium alloy increased with increasing pressure, the degree of hardening of the material was reduced and the residual compressive stress of the treated surface increased.

Sun et al. [29] tested the machined surface integrity after turning TB6 titanium alloy. The influence of machining parameters and technological system vibrations on the roughness and defects of the treated surface was analyzed. It was found that feed rate affects surface roughness, residual stresses and microhardness significantly. The Ra parameter increased from 0.2 to 0.4 μm, the residual stress increased by 10%, and the peak position increased from 50 to 70 μm when the feed rate ranged from 0.05 to 0.1 mm/rev. In turn, the vibrations frequency of the technological system was similar to the frequency of surface roughness profile changes.

Mishra et al. [30] studied the mechanical properties of Ti6Al4V ELI titanium alloy during turning with a coated Wiper insert under fluid cooling conditions. It was found that the cutting speed (percentage effect up to ~63%) is an important parameter affecting the wear of the insert, followed by the depth of cut (percentage effect up to ~28%). It was also found that the feed rate is the main parameter influencing the surface roughness (percentage effect up to ~71%), whereas the depth of cut has percentage influence up to ~20% and the cutting speed up to ~8%, i.e. they are less important.

Rahman et al. [31] machined Ti6Al4V ELI titanium alloy under cooling conditions using the MQCL (minimum quantity cooling and lubrication) method with the addition of nanofluid. The results were related to dry machining and MQCL on the basis of mineral oil. Nanofluids showed a reduction of tool wear and prevented the tool abrasion. The authors noticed traces of tribofilm and the effect of nano-polishing of the surface machined, which contributed to obtaining lower values of the surface roughness parameter Ra and decreased the number of surface defects of the material.

Yi et al. [32] analyzed the formation of chips, temperature and surface roughness in the process of Ti6Al4V alloy turning under conditions of a new type of cooling with the use of graphene-based nanofluids. The influence of pressure and concentration of coolant as well as cutting speed, feed rate and cutting depth were investigated. It was shown that the surface roughness was reduced by ~34% compared to conventional coolant. The cutting temperature was reduced by ~27 °C, ~30 °C and ~32 °C with the lower cutting speed, lower feed rate and higher coolant pressure respectively. The cutting temperature decreased by ~50 °C when the pressure of the nanofluids flow was 10 bar.

Shaping the surface texture when turning of titanium alloys arouses a lot of interest, and difficulties of processing these materials have resulted in the conduct of studies these processes modeling. Schultheiss et al. [33] modeled the influence of chip thickness on surface roughness (Ra parameter) during turning of Ti6Al4V alloy. D'Mello et al. [34] modeled the surface roughness (Ra and Rt parameters) taking into account the cutting speed, feed rate and cutting depth, as well as uncontrolled wear parameters and tool vibrations in the process of Ti6Al4V titanium alloy turning under dry conditions.

The worse the surface integrity is, the lower is the fatigue strength, which is responsible for nearly 90% of failures in operation due to mechanical reasons [35–38].

The unfavourable properties of the machined surface of Ti6Al4V implants remain a base limitation of fast and stable bone tissue integration. Therefore, in recent years we have been looking for a way to reduce the surface roughness of the titanium Ti6Al4V alloy.

Therefore, the aim of the study was to determine the influence of cutting parameters on the surface texture parameters of Ti6Al4V titanium alloy when finish turning under dry and wet cooling conditions.

2   Experimental Procedure

Ti6Al4V titanium alloy with chemical composition (%, according to the ISO 5832-3): O < 0.20, V = 3.5, Al = 5.5, Fe < 0.30, H < 0.0015, C < 0.08, N < 0.05, Ti = the rest was machined. Its mechanical properties are: modulus of elasticity 110–114 [GPa], tensile strength 960–970 [MPa], yield strength 850–900 [MPa], and fatigue strength 620–725 [MPa].

The research was carried out with the use of CTX 510 type CNC lathe centre (Fig. 1). A turning tool with CoroTurn SDJCR 2525 M 11 holder and CoroTurn DCMX 11 T3 04-WM 1115 insert was used and $(Ti,Al)N + (Al,Cr)_2O_3$ coating was deposited on inserts by the PVD method. Angles of the cutting wedge were: tool cutting edge angle $K_r = 93°$, rake angle $\gamma = 18°$, clearance angle $\alpha = 7°$, corner radius $r_\varepsilon = 0.4$ mm, cut width $b_{\gamma n} = 0.1$ mm.

Fig. 1. CTX 510 type CNC lathe centre with a test workpiece.

The investigations were carried out in the range of cutting speeds 40–120 m/min and feed rates 0.05–0.4 mm/rev. The stable cutting depth of 0.5 mm was used, which corresponds with the conditions of finishing machining.

The machining was carried out without the use of cooling lubricant (dry cutting) and with the use of emulsion based on water and the emulsifying oil Castrol Alusol SL 51 XBB emulsifying with 7% working concentration (wet cutting).

The research plan was based on the PSI (Parameter Space Investigation) method, which allows planning an experiment while minimizing the points of experience. The test points were placed in fixed locations in a sequential manner. The sequence consists in placing the points in a multidimensional space in such way that the points of their projection are located respectively on the X_1 and X_2 axes and are at equal distances between themselves (Fig. 2) [39]. Based on the algorithm presented in [37], the coordinates of the test points in the range $X_{min} = 0$ and $X_{max} = 1$ were calculated, and they are shown in Table 1. The PSI method was successfully applied previously in the machining process studies [40, 41].

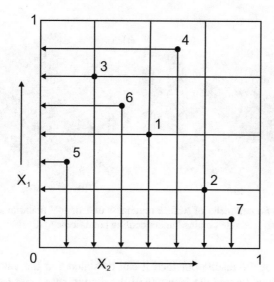

Fig. 2. Distribution of test points on X_1 and X_2 axes in accordance with the PSI method.

Table 1. Coordinates of PSI test points.

Variables	1	2	3	4	5	6	7
X_1	0.5000	0.2500	0.7500	0.8750	0.3750	0.6250	0.1250
X_2	0.5000	0.7500	0.2500	0.6250	0.1250	0.3750	0.8750

The number of test points on each axis (7 points) is satisfactory for statistical calculations. The analysis of the measurement results of the tests was performed using the Statistica 13 software.

The surface texture after turning of Ti6Al4V titanium alloy was measured with the Keyence VHX-6000 digital microscope.

The results were obtained for the surface roughness parameter Ra and have been statistically analyzed. According to the PN-EN ISO 4287:1999 standard Ra is a deviation of the arithmetic mean of the roughness profile, and in the industry it is interpreted as a quality requirement related to the geometrical features of the product.

3   Results

The graphs characterizing the changes in the *Ra* roughness parameter depending on the variable cutting speeds and feed rate under dry and wet cooling conditions are shown in Fig. 3.

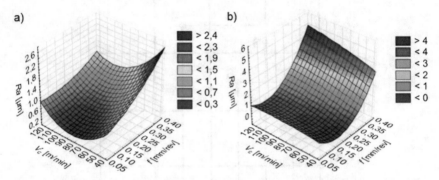

Fig. 3. Changes in *Ra* parameter of surface roughness of Ti6Al4V titanium alloy depending on the cutting speed and feed rate obtained under cooling conditions: a) dry, b) wet.

On the basis of the results obtained it can be stated that the value of the surface roughness parameter *Ra* and the intensity of the cutting speed and feed rate influence when turning of Ti6Al4V titanium alloy under dry conditions decreases in comparison with the wet machining. It was found that when dry machining lower values of roughness parameters *Ra* were obtained in the range of lower feed rates and average cutting speeds, whereas under wet conditions it was observed in the range of lower feed rates and the whole range of cutting speed. The differences indicated may be caused by simultaneous changes of plastic-elastic stresses and heat transfer conditions in the cutting zone. These changes affect the type and shape of the chip and depend on a number of cutting factors, including cutting parameters and cooling conditions.

Figure 4 presents surface roughness profiles after turning of Ti6Al4V titanium alloy in PSI test points, where minimum and maximum values of surface roughness parameters were obtained under dry and wet cooling conditions. The parameters of cutting $V_c = $ ~81 m/min and $f = 0.225$ mm/rev in case of dry machining and $V_c = $ ~103 m/min and $f = 0.137$ mm/rev under wet conditions, where lower values of the roughness parameter *Ra* were obtained, are worth mentioning. However, under both cooling conditions higher values of the *Ra* roughness parameter were obtained by turning with $V_c = $ ~48 m/min and $f = 0.356$ mm/rev.

Lower values of surface roughness parameters under dry and wet turning were observed at higher cutting speeds in the range studied. It is caused by temperature increase in the cutting zone and plasticization of the workpiece material at increased speed. Higher values of surface roughness parameters in both machining conditions were observed with the use of high feed rates, which is a commonly known phenomenon:

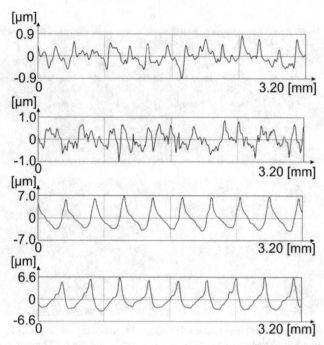

Fig. 4. Surface roughness profiles obtained under selected conditions: a) dry conditions with $V_c = {\sim}81$ m/min and $f = 0.225$ mm/rev, b) wet conditions with $V_c = {\sim}103$ m/min and $f = 0.137$ mm/rev, c) dry conditions with $V_c = {\sim}48$ m/min and $f = 0.356$ mm/rev, d) wet conditions with $V_c = {\sim}48$ m/min and $f = 0.356$ mm/rev.

with the increase of feed rate, the share of geometric-kinetic representation of the cutting wedge in the shaping of roughness increasing [42].

Figure 5 shows Ti6Al4V titanium alloy surface images after finish turning under dry and wet cooling conditions, in which the best and worst values of surface roughness parameters were obtained.

Figure 5a shows a surface image of a titanium alloy Ti6Al4V under dry conditions with $V_c = {\sim}81$ m/min and $f = 0.225$ mm/rev for which $Ra = 0.223$ μm and Fig. 5b shows this for wet turning with $V_c = {\sim}103$ m/min and $f = 0.137$ mm/rev for which Ra was 0.284 μm. For the cooling conditions studied, lower values of the Ra parameter were obtained at various test points. Although these values are similar to each other, a higher number of surface defects can be observed when dry machining of Ti6Al4V titanium alloy. These defects are caused by the adhesion of the chip chips to the treated surface. When emulsion was used, these particles were rinsed off the treated surface. Figure 5c shows the surface image observed when dry turning of Ti6Al4V titanium alloy with $V_c = {\sim}48$ m/min and $f = 0.356$ mm/rev with $Ra = {\sim}2.389$ μm. Such a surface is characteristic of the conditions of the highest Ra value for dry machining. In this case, adhesion of fine chips to the treated surface was observed on the material

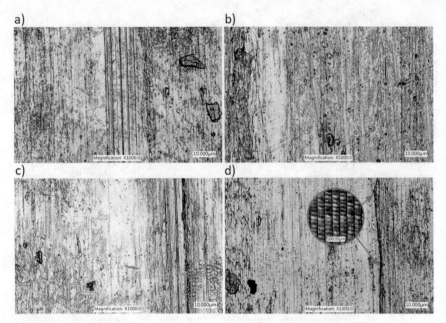

Fig. 5. 2D surface images obtained under selected conditions: a) dry cutting with $V_c = {\sim}81$ m/min and $f = 0.225$ mm/rev, b) wet cutting with $V_c = {\sim}103$ m/min and $f = 0.137$ mm/rev, c) dry cutting with $V_c = {\sim}48$ m/min and $f = 0.356$ mm/rev, d) wet cutting with $V_c = {\sim}114$ m/min and $f = 0.269$ mm/rev.

surface. Figure 5d shows the surface image obtained when wet machining with $V_c = {\sim}114$ m/min and $f = 0.269$ mm/rev, for which $Ra = 1.446$ µm. For these conditions, there are adhesive bonding of fine chips with the machined surface and very significant vibrations are visible (the image in the red frame). Despite the presence of emulsion, the adhesive bonds have not been washed out, which can be a result of the high temperature in the cutting zone and the increase of contact stresses.

Relative changes of the Ra of surface roughness parameter in the PSI test points, depending on cooling conditions, were also analyzed. In Fig. 6 the average percentage changes of Ra parameter obtained after finish turning of Ti6Al4V titanium alloy under dry and wet conditions depending on the cutting speed and feed values are presented.

Reduction of the Ra parameter value of surface roughness was obtained under dry machining in the range of 3% to 50% compared to wet machining. On the other hand, the increase was obtained in the range from 11% to 137% for wet turning in comparison with dry turning. The boundary of changes in the surface roughness parameters in Fig. 6 was marked with a black dotted line. According to this boundary, it can be stated that the decrease in the Ra parameter of the surface roughness occurs at lower and medium cutting speeds in all feed range studied, while the higher Ra values are observed at medium and higher cutting speeds.

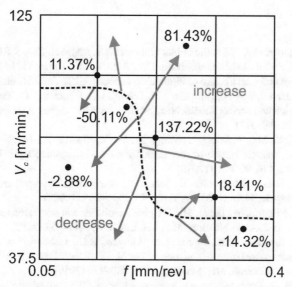

Fig. 6. Average percentage changes of *Ra* parameter values of surface roughness obtained at PSI test points under wet turning in comparison with dry turning.

4   Conclusions

In this paper the formation of the surface texture of Ti6Al4V titanium alloy after finish turning under dry and wet cooling conditions was investigated. Variable cutting speeds and feed rates with a constant cutting depth were applied. The *Ra* parameter of surface roughness was analyzed. On the basis of the test results obtained it was found that:

1. Under both cooling conditions, dry cutting and wet cutting, lower feed rates reduce the *Ra* value of the surface roughness. On the other hand, the effect of the cutting speed is insignificant.
2. When turning under dry cooling conditions, both the *Ra* value of the surface roughness and the intensity of the impact of the cutting speed and feed rate are reduced compared to wet machining.
3. During dry machining, a lower *Ra* value was obtained for surface roughness, however, a higher number of defects were observed on the machined surface compared to wet machining.
4. Under dry machining compared to wet machining, a reduction of *Ra* parameters of surface roughness of up to 50% was achieved.
5. More favorable surface roughness features were obtained when dry machining was carried out using middle cutting speeds (~80 m/min) and feed rates (~0.2 mm/rev).

Acknowledgments. The authors gratefully acknowledge the financial support from the program of the Polish Minister of Science and Higher Education under the name "Regional Initiative of Excellence" in 2019–2022, project no. 003/RID/2018/19, funding amount 11 936 596.10 PLN".

References

1. Chen, Q., Thouas, G.A.: Metallic implant biomaterials. Mat. Sci. Eng. R **87**, 1–57 (2015)
2. Ramsden, J.J., Allen, D.M., Stephenson, D.J., Alcock, J.R., Peggs, G.N., Fuller, G., Goch, G.: The design and manufacture of biomedical surfaces. CIRP Ann. **56**, 687–711 (2007)
3. Asri, R.I.M., Harun, W.S.W., Samykano, M., Lah, N.A.C., Ghani, S.A.C., Tarlochan, F., Raza, M.R.: Corrosion and surface modification on biocompatible metals: a review. Mat. Sci. Eng. C **77**, 1261–1274 (2017)
4. Zhou, G., Bi, Y., Ma, Y., Wang, L., Wang, X., Yu, Y., Mutzke, A.: Large current ion beam polishing and characterization of mechanically finished titanium alloy (Ti6Al4V) surface. Appl. Surf. Sci. **476**, 905–913 (2019)
5. Machara, K., Doi, T., Matsushita, Y., Susaki, S.: Attenuation of ischemia/reperfusion injury in rats by a caspase inhibitor. Mater. Trans. **43**, 2936–2942 (2002)
6. Boehlert, C., Niinomi, M., Ikedu, M.: The phase evolution and microstructural stability of an orthorhombic Ti-23Al-27 Nb alloy. Mat. Sci. Eng. C **25**, 247–252 (2005)
7. Costa, B.C., Tokuhara, C.K., Rocha, L.A., Oliveira, R.C., Lisboa-Filho, P.N., Costa, P.J.: Vanadium ionic species from degradation of Ti-6Al-4V metallic implants: in vitro cytotoxicity and speciation evaluation. Mat. Sci. Eng. C **96**, 730–739 (2019)
8. Manivasagam, G., Dhinasekaran, D., Rajamanickam, A.: Biomedical implants: corrosion and its prevention - a review. Recent Patents Corros. Sci. **2**, 40–54 (2010)
9. Li, P., Warner, D.H., Fatemi, A., Phan, N.: Critical assessment of the fatigue performance of additively manufactured Ti–6Al–4V and perspective for future research. Int. J. Fatigue **85**, 130–143 (2016)
10. Nicoletto, G., Konečná, R., Frkáň, M., Riva, E.: Surface roughness and directional fatigue behavior of as-built EBM and DMLS Ti6Al4V. Int. J. Fatigue **116**, 140–148 (2018)
11. Geetha, M., Singh, A.K., Asokamani, R., Gogia, A.K.: Ti based biomaterials, the ultimate choice for orthopaedic implants – a review. Prog. Mater Sci. **54**, 397–425 (2009)
12. Liang, L., Liu, X., Li, X., Li, Y.Y.: Wear mechanisms of WC–10Ni3Al carbide tool in dry turning of Ti6Al4V. IJRMHM **48**, 272–285 (2015)
13. Michailidis, N.: Variations in the cutting performance of PVD-coated tools in milling Ti6Al4V, explained through temperature-dependent coating properties. Surf. Coat. Tech. **304**, 325–329 (2016)
14. Sartori, S., Ghiotti, A., Bruschi, S.: Solid lubricant-assisted minimum quantity lubrication and cooling strategies to improve Ti6Al4V machinability in finishing Turning. Tribol. Int. **118**, 287–294 (2018)
15. Batista, M., Salguero, J., Gomez-Parra, A., Fernández-Vidal, S., Marcos, M.: SOM based methodology for evaluating shrinkage parameter of the chip developed in titanium dry turning process. Procedia CIRP **8**, 534–539 (2013)
16. Ahmad Yasir, M.S., Che Hassan, C.H., Jaharah, A.G., Nagi, H.E., Yanuar, B., Gusri, A.I.: Machinalibilty of Ti-6Al-4 V under dry and near dry condition using carbide tools. IJIMSE **2**, 1–9 (2009)
17. Bai, W., Sun, R., Roy, A., Silberschmidt, V.V.: Improved analytical prediction of chip formation in orthogonal cutting of titanium alloy Ti6Al4V. Int. J. Mech. Sci. **133**, 357–367 (2017)
18. Childs, T.H.C., Arrazola, P.J., Aristimuno, P., Garay, A., Sacristan, I., Mater, J.: Ti6Al4V metal cutting chip formation experiments and modeling over a wide range of cutting speeds. Process. Tech. **255**, 898–913 (2018)
19. Jianxin, D., Yousheng, L., Wenlong, S.: Diffusion wear in dry cutting of Ti–6Al–4V with WC/Co carbide tools. Wear **265**, 1776–1783 (2008)

20. Maruda, R.W., Krolczyk, G.M., Nieslony, P., Wojciechowski, S.: Structura and microhardness changes after turning of the AISI 1045 steel for minimum quantity cooling lubrication. J. Mater. Eng. Perform. **26**, 431–438 (2017)

21. Maruda, R.W., Krolczyk, G.M., Wojciechowski, S., Zak, K., Habrat, W., Nieslony, P.: Effects of extreme pressure and anti-wear additives on surface topography and tool wear during MQCL turning of AISI 1045 steel. J. Mech. Sci. Tech. **32**, 1585–1591 (2018)

22. Sartori, S., Bordin, A., Ghiotti, A., Bruschi, S.: Analysis of the surface integrity in cryogenic turning of Ti6Al4V produced by direct melting laser sintering. Procedia CIRP **45**, 123–126 (2016)

23. Bordin, A., Sartori, S., Bruschi, S., Ghiotti, A.: Experimental investigation on the feasibility of dry and cryogenic machining as sustainable strategies when turning Ti6Al4V produced by additive manufacturing. J. Clean. Prod. **142**, 4142–4151 (2017)

24. Mia, M., Gupta, M.K., Lozano, J.A., Carou, D., Pimenov, D.Y., Królczyk, G., Khan, A.M., Dhar, N.R.: Multi-objective optimization and life cycle assessment of eco-friendly cryogenic N_2 assisted turning of Ti-6Al-4V. J. Clean. Prod. **210**, 121–133 (2019)

25. Jamil, M., Khan, A.M., Hegab, H., Gong, L., Mia, M., Gupta, M.K., He, N.: Effects of hybrid Al2O3-CNT nanofluids and cryogenic cooling on machining of Ti–6Al–4V. Int. J. Adv. Manuf. Technol. **102**, 3895–3909 (2019)

26. Mia, M., Dhar, N.R.: Influence of single and dual cryogenic jets on machinability characteristics in turning of Ti-6Al-4V. Proc. Inst. Mech. Eng. Part B-J. Eng. Manuf. **233**, 711–726 (2017)

27. Liang, X., Liu, Z., Tao, G., Wang, B., Ren, X.: Investigation of surface topography and its deterioration resulting from tool wear evolution when dry turning of titanium alloy Ti6Al4V. Tribol. Int. **135**, 130–142 (2019)

28. Liang, X., Liu, Z., Liu, W., Wang, B., Tao, G.: Surface integrity analysis for high-pressure jet assisted machined Ti6Al4V considering pressures and injection positions. J. Manuf. Process. **40**, 149–159 (2019)

29. Sun, J., Huang, S., Wang, T., Chen, W.: Research on surface integrity of turning titanium alloy TB6. Procedia CIRP **71**, 484–489 (2018)

30. Mishra, R.R., Kumar, R., Sahoo, A.K., Panda, A.: Machinability behaviour of biocompatible Ti-6Al-4V ELI titanium alloy under flood cooling environment. Mater. Today: Proc. (2019, in Press)

31. Rahman, S.S., Ashraf, M.Z.I., Amin, A.N., Bashar, M.S., Ashik, M.F.K., Kamruzzamann, M.: Turning nanofluids for improved lubrication performance in turning biomedical grade titanium alloy. J. Clean. Prod. **206**, 180–196 (2019)

32. Yi, S., Mo, J., Ding, S.: Experimental investigation on the performance and mechanism of graphene oxide nanofluids in turning Ti-6Al-4V. J. Manuf. Process. **43**, 164–174 (2019)

33. Schultheiss, F., Hägglund, S., Bushlya, V., Zhou, J., Ståhl, J.E.: Influence of the minimum chip thickness on the obtained surface roughness during turning operations. Procedia CIRP **13**, 67–71 (2014)

34. D'Mello, G., Pai, P.S., Shetty, R.P.: Surface roughness modeling in high speed turning of Ti6Al4V – artificial neural network approach. Mater. Today: Proc. **4**, 7654–7664 (2017)

35. Craig, R.G., Powers, J.M.: Cast and wrought base metal alloys. In: Restorative Dental Materials, 11th edn. Mosby Inc., St. Louis (2002)

36. Zavanelli, R.A., Pessanna Henriques, G.E., Ferreira, I., De Almeida Rollo, J.M.: Corrosion-fatigue life of commercially pure titanium and Ti-6Al-4V alloys in different storage environments. J. Prosthet. Dent. **84**, 274–279 (2000)

37. Vallittu, P.K., Kokkonen, M.: Deflection fatigue of cobalt-chromium, titanium and gold alloy cast denture clasp. J. Prosthet. Dent. **74**, 412–419 (1995)

38. Guilherme, A.S., Henriques, G.E., Zavanelli, R.A., Mesquita, M.F.: Surface roughness and fatigue performance of commercially pure titanium and Ti-6Al-4V alloy after different polishing protocols. J. Prosthet. Dent. **93**, 378–385 (2005)
39. Trent, E.M., Wright, P.K.: Metal Cutting, 4th edn. Butterworth-Heinemann, Woburn (2000)
40. Maruda, R.W., Krolczyk, G.M., Niesłony, P., Krolczyk, J.B., Legutko, S.: Chip formation zone analysis during the turning of austenitic stainless steel 316L under MQCL cooling condition. Procedia Eng. **149**, 297–304 (2016)
41. Maruda, R.W., Legutko, S., Krolczyk, G.M., Hloch, S., Michalski, M.: An influence of active additives on the formation of selected indicators of the condition of the X10CrNi18-8 stainless steel surface layer in MQCL conditions. Int. J. Surf. Sci. Eng. **9**, 452–465 (2015)
42. Habrat, W.: Analysis and modeling of the finish turning of titanium and its alloys. Oficyna Wydawnicza Politechniki Rzeszowskej, Rzeszów (2019)

Surface Roughness Analysis After Milling Aluminum-Ceramic Composite

Natalia Znojkiewicz$^{(\boxtimes)}$, Paweł Twardowski, Jakub Czyżycki, and Marek Madajewski

Faculty of Mechanical Engineering and Management, Poznan University of Technology, Poznań, Poland
natalia.w.znojkiewicz@doctorate.put.poznan.pl

Abstract. The paper presents the results of surface roughness measurements after milling with an end mill of an aluminum-ceramic composite. The tests were carried out on a DMC 70V DECKEL MAHO machining center, and the cutting tools were CBN and PCD cutters. The tests were carried out with variable cutting speeds and variable feeds with ant without presence of cutting fluids. For surface roughness analysis the *Ra, Rq, Rz, Rt* and *Sa, Sq, Sz, St* parameters were selected. The analysis of the influence of the variation in cutting parameters (cutting speed v_c and feed f) on the roughness of the machined surface was carried out under the same conditions for both tool materials. In the further part of the paper, surface roughness models were analyzed whether they include kinematic and geometric parameters in the workpiece. The kinematic-geometrical model for the milling process assumes that the main factor determining the surface roughness of the machined surface is the feed per tooth and the diameter of the milling cutter. Literature reports as well as own research allow to conclude that during milling process, feed per revolution is decisive factor for surface roughness, which is contrary to theoretical model where only feed per tooth has effect on surface roughness parameters. These models were verified with experimental data and their effectiveness was assessed on this basis.

Keywords: Milling · Aluminum-ceramic composite · Surface roughness prediction models · PCD · CBN

1 Introduction

Metal matrix composites (MMCs) have become the leading material among composite materials, and in particular those reinforced with aluminum particles have gained considerable weight due to their excellent technical properties. These materials are known to be difficult to cut, due to the hardness and abrasiveness of reinforcing SiC silicon carbide particles [1].

Composites quickly replace conventional materials in a variety of engineering applications, such as the aerospace and automotive industries [1, 2]. The most popular reinforcements are silicon carbide (SiC) and aluminum oxide (Al_2O_3). As the matrix phase, aluminum, titanium and magnesium alloys are commonly used [1, 3].

© Springer Nature Switzerland AG 2020
G. M. Królczyk et al. (Eds.): IMM 2019, LNME, pp. 45–54, 2020.
https://doi.org/10.1007/978-3-030-49910-5_5

The main problem during machining of MMC is the significant wear of tools caused by very hard reinforcements [1]. Milling is currently one of the subtractive techniques used in the engineering industry for manufacturing.

The main problem faced by the technologist when designing the milling process is to achieve previously defined product quality for a given equipment, cost estimation and time limits. Unfortunately, for some product quality characteristics such as the geometry of the surface (GPS), it is difficult to be sure that these requirements will be met, because the GPS generation process during milling is very complex and influenced by many factors.

To overcome these problems throughout the world, work is underway to develop models that attempt to simulate conditions during milling, describe the causes and results of the relationship between the various factors that ultimately affect the final quality of the product.

The ability to plan the surface quality is very important in production due to meeting the requirements for parts machined by various methods. The problem of modeling surface roughness during milling of difficult-to-cut materials has been undertaken by researchers for many years. For example, Baek et al. [4] proposed an extended surface roughness model taking into account the milling parameters, tool geometry and relative movements between the workpiece and the cutting tool. Zhenyu et al. [5] established a similar dynamic model dedicated to high speed milling using a straight edge tool with square inserts. Peigne et al. [6] formulated a rigid milling cutter - a flexible dynamic model of the object to assess surface roughness during cylindrical milling. Buj-Corral et al. [7] formulated the surface roughness model for end milling, taking into account the radius of the tool's corner, the feed per tooth, the angle of run-out and the angle of inclination of the main cutting edge.

The article [8] introduces the method and procedure of modeling based on the definition of the relationship between the calculated - theoretical and measured - the real value of the surface roughness of the machined surfaces. The authors conducted experiments to determine the actual roughness values in which the surface roughness was tested with various technological parameters. Finally, the calculated relationships between theoretical and real values are presented, which allows predicting the expected surface roughness milled given conditions.

Work [9] was focused on the prediction of surface roughness after milling of hardened steels. The developed models included the kinematic-geometric projections, as well as the instant tool displacements. The paper [10] presented an analysis of the relationship between instantaneous tool displacements and the surface roughness created during the milling of a spherical surface with an inclination in the direction of the tool axis. The experiments were carried out on hardened, low-carbon alloy steel. Research shows that the value of the tool overhang significantly affects the mechanisms of surface roughness generation during finishing milling with a spherical cutter. In the case of milling with a rigid tool ($l = 35$ mm), the surface roughness is strongly correlated with the kinematic-geometric model as well as geometrical errors of the machine tool. Nevertheless, in the case of milling with a slender tool ($l = 85$ mm), the surface roughness is mainly influenced by the dynamic deflection of the tool caused by the milling forces.

Researchers also conduct research showing the current state-of-the-art regarding surface roughness models and their application in practice.

The authors [11] have noticed that despite the large number of scientific studies on empirical surface roughness models, there is no methodology used in industry to model and accurately adapt surface roughness in machining operations. Each change in the process, regarding an initial condition in which the experiments were carried out entails an additional estimation error which makes it difficult to use the model in the current process. Authors [11] in their research investigated the portability of empirical models to predict surface roughness in face milling operations. As a problem with portability, it refers to how the correct surface roughness model obtained from theoretical/experimental data under specific conditions reduces its performance when used in a different environment. This work gives some hints on the future design of surface roughness models.

The purpose of the article [12] was to review the current state of knowledge in the area of surface roughness and factors influencing the surface roughness. The work also discusses the challenges and opportunities facing industry and academia.

The article presents results regarding surface roughness analysis after milling aluminum-ceramic composite. In addition, a comparison of current roughness models with the obtained real values was made.

2 Experimental Details

The tests were carried out on a three-axis milling center from the DECKEL MAHO DMU 60monoBLOCK model (Fig. 1) with a maximum rotational speed of the electro-spindle of $n = 24\,000$ rev/min.

The research involved two monolithic end mills (Table 1) with diameter $D = 10$ mm and different number of teeth. The CBN cutter was without an anti-wear coating, while the PCD cutter was carbide with a diamond coating.

Fig. 1. Milling center from DECKEL MAHO

Table 1. Cutters used in tests

CBN Cutter	PCD Cutter
z = 2	z = 6

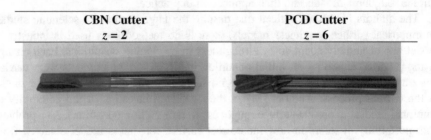

In the research the aluminum-ceramic composite from DURALCAN company was employed. It consists of an aluminum alloy matrix, reinforced with SiC molecules containing up to 10% of the SiC particles' volume.

The chemical composition of the sample is shown in Table 2.

Table 2. The chemical composition of the DURALCAN composite

Chemical composition, % by weight							
Si	Fe	Cu	Mn	Mg	Zn	Ti	Al
8.88	0.07	0.001	0.002	0.62	0.002	0.10	Rest

The tests were carried out with and without the use of cutting fluids. The tests included measuring the roughness of the surface after the milling process of the aluminum-ceramic composite with varying milling parameters (Table 3). Although the cutters had a different number of blades, the feed per revolution f was the same for the two milling cutters tested. This is due to the fact that the feed per revolution f has a much greater effect on the surface roughness than on the feed on the blade f_z. Hence, maintaining the constant feed value per revolution f.

In the further part of the work, surface roughness models were analyzed, which reflect the kinematic and geometric parameters. These models were verified with experimental data and their effectiveness was assessed.

Table 3. The milling parameters used

Tool material	D [mm]	z	a_e [mm]	a_p [mm]	n [rev/min]	v_c [m/min]	f [mm/rev]	f_z [mm/tooth]
PCD	10	6	0.3	10	15916	500	0.01 ÷ 0.5	0.0017 ÷ 0.083
					1591 ÷ 22273	50 ÷ 700	0.1	0.02
CBN		2			15916	500	0.01 ÷ 0.5	0.0049 ÷ 0.25
					1591 ÷ 22273	50 ÷ 700	0.1	0.05

3   Results and Discussion

During the tests, milling was carried out without cutting fluids, so-called "dry" series. During the dry tests, the spatial 3D parameters of roughness were analyzed: *Sa* – *arithmetic mean roughness deviation,* *Sz* – *ten-point height of unevenness of the surface,* *St* – *height of surface irregularities* and *Sq* – *square mean roughness deviation.*

Figure 2 presents the effect of cutting speed v_c on the surface roughness parameter *St* and surface topography for selected points in the chart.

From the graph below it can be seen that the impact of v_c on the roughness for the tools used is non-monotonic. For PCD tool there appeared self-excited vibrations which caused a several-fold increase in surface roughness parameters. Self-excited vibrations were manifested by the occurrence of high-frequency noise, which adversely affected the roughness of the surface.

Figure 3 shows the effect of feed per tooth f_z on the surface roughness parameter *Sq*. The lowest values were obtained for CBN tool, whereas for PCD, the feed effect was most pronounced.

In the next part of the research, milling was carried out using a cutting liquids, so-called "wet" series. During the wet tests, 2D surface roughness parameters were analyzed: *Ra* – *arithmetic mean of the ordinates of the roughness profile,* *Rz* – *highest roughness profile height,* *Rt* – *total height of roughness profile* and *Rq* – *square average of the ordinates of the roughness profile.*

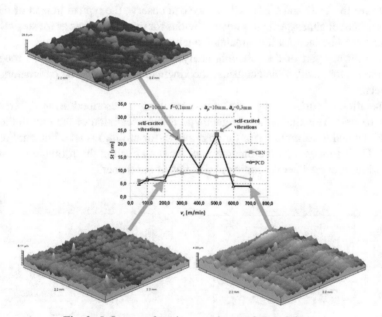

Fig. 2. Influence of cutting speed on surface roughness

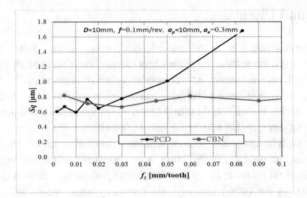

Fig. 3. Influence of feed values on surface roughness

Figures 4 and 5 show the influence of the cutting speed on the *Ra* parameter of the analyzed cutters.

Based on the graphs below, it is clearly visible that there is no unequivocal influence of v_c on the analyzed roughness parameter. In the conducted scope of research, v_c has no effect on surface roughness.

Figures 6 and 7 show the influence of the feed per tooth value on the *Ra* parameter of the analyzed cutters.

Both for the PCD and CBN cutters, one can observe the typical impact of the feed on the surface roughness, which is not as obvious for the composites as for steel. Similar relations were obtained for the remaining analyzed parameters.

In the further part of the research, analysis and verification of surface roughness models were presented, which consider the kinematic and geometric parameters in the workpiece.

Generation of surface roughness after machining is described using different theoretical models. The kinematic and geometrical representation of the tool in the work material defined in formula 1 (for turning) and 2 (for milling) is well known. From the formula (1) it follows that the predominant factors affecting the formation of surface irregularities are: feed per revolution and radius of the corner.

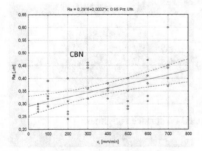

Fig. 4. Influence of the cutting speed value on the *Ra* parameter, for the CBN cutter.

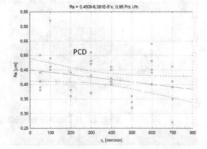

Fig. 5. Influence of the cutting speed value on the *Ra* parameter, for the PCD cutter.

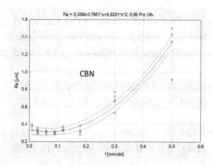

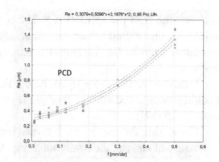

Fig. 6. Influence of the feed f on the Ra parameter, for the CBN cutter.

Fig. 7. Influence of the feed f on the Ra parameter, for the PCD cutter.

$$R_{zth} = \frac{f^2}{8 \cdot r_\varepsilon} \tag{1}$$

where: f – feed per revolution [mm/rev]
r_ε – corner radius [mm]

$$R_{zth} = \frac{f_z^2}{4d} \tag{2}$$

where: f_z – feed per tooth [mm/tooth]
d – diameter of the cutter [mm].

Unfortunately, this model works only for large feed values, which in practice are not used for precise machining. The Brammertz model [13] is definitely more accurate, taking into account the minimum uncut chip thickness - h_{min} (3).

$$R_{zth} = \frac{f^2}{8 \cdot r_\varepsilon} + \frac{h_{min}}{2}\left(1 + \frac{r_\varepsilon \cdot h_{min}}{2}\right) \tag{3}$$

In literature there is also an extended version of the formula 2 taking into account the elastic recovery of the machined layer [14] (4).

$$R_{zth} = \frac{f^2}{8 \cdot r_\varepsilon} + \frac{h_{min}}{2}\left(1 + \frac{r_\varepsilon \cdot h_{min}}{2}\right) + k_1 \cdot r_n \cdot \frac{H}{E} \cdot k_2 \tag{4}$$

where: k_1, k_2 – coefficients related elastic springback of workpiece
H – Vickers hardness
E – Young's modulus.

In Figs. 8 and 9, a preliminary verification of the kinematic-geometric model and the Brammertz model for milling was made.

The red line is the second degree polynomial for experimental data. The blue line is a model taking into account the feed per tooth f_z, while the green line is a model that includes the feed per revolution f. The black line is a Brammertz model including the minimum uncut chip thickness - h_{min}, and feed per tooth f_z, while the gray line is a Brammertz model that includes the minimum uncut chip thickness - h_{min} and feed per revolution f.

It can be seen from the figure below that the kinematic-geometric model for milling should take into account the feed per revolution (this is also confirmed by the literature data [6, 15–17]).

Comparing the differences between the real and theoretical values for the kinematic-geometric model and the Brammertz (Fig. 9), it can be noted that the smaller the feed, the larger the differences. Therefore, the classic kinematical-geometric model is not suitable for modeling surface roughness for small feed values - this is its basic limitation. The Brammertz model is definitely more accurate.

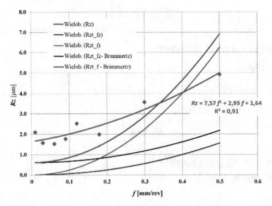

Fig. 8. Verification of kinematic-geometric and Brammertz models taking into account the minimum uncut chip thickness.

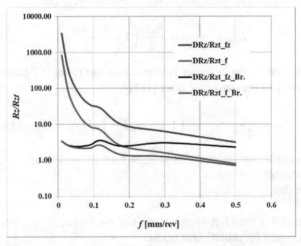

Fig. 9. The difference between the real and theoretical values for the kinematic-geometric model and the Brammertz.

4 Conclusions

In the studied range of parameter variability, it was shown that:

- The feed has a decisive influence on surface roughness parameters, which is not obvious in the case of aluminum-ceramic composites.
- The cutting speed, in turn, does not have a significant impact on the surface roughness, unless self-excited vibrations deteriorate the roughness parameters. In such situations, the spindle speed must be changed.
- The kinematic-geometric model is not suitable for modeling surface roughness for small feed values - this is its basic limitation.
- A much more accurate model for surface roughness modeling is the Brammertz model taking into account the minimum uncut chip thickness.
- In further studies, a model should be developed that takes into account the hard reinforcements presented in the composite, which will allow the modeling of the surface roughness of the aluminum-ceramic composite.

References

1. Seeman, M., Ganesan, G., Karthikeyan, R., Velayudham, A.: Study on tool wear and surface roughness in machining of particulate aluminum metal matrix composite-response surface methodology approach. Int. J. Adv. Manuf. Technol. **4**, 613–624 (2010)
2. Ding, X., Liew, W.Y.H., Liu, X.D.: Evaluation of machining performance of MMC with PCBN and PCD tools. Wear **259**, 1225–1234 (2005)
3. Quan, Y., Ye, B.: The effect of machining on the surface properties of SiC/Al composites. J. Mater. Process. Technol. **138**, 464–467 (2003)
4. Kyun, B.D., Jo, K.T., Sool, K.H.: A dynamic surface roughness model for face milling. Precis. Eng. **20**, 171–178 (1997)
5. Zhenyu, S., Luning, L., Zhanqiang, L.: Influence of dynamic effects on surface roughness for face milling process. Int. J. Adv. Manuf. Technol. **80**, 1823–1831 (2015)
6. Peigne, G., Paris, H., Brissaud, D., Gouskov, A.: Impact of the cutting dynamics of small radial immersion milling operations on machined surface roughness. Int. J. Mach. Tools Manuf **44**, 1133–1142 (2004)
7. Buj-Corral, I., Vivancos-Calvet, J., González-Rojas, H.: Influence of feed, eccentricity and helix angle on topography obtained in side milling processes. Int. J. Mach. Tools Manuf **51**, 889–897 (2011)
8. Felhő, C., Karpuschewski, B., Kundrák, J.: Surface roughness modelling in face milling. Procedia CIRP **31**, 136–141 (2015)
9. Twardowski, P., Wojciechowski, S., Wieczorowski, M., Mathia, T.: Surface roughness analysis of hardened steel after high-speed milling. Scanning **33**, 1–10 (2011)
10. Wojciechowski, S., Wiackiewicz, M., Krolczyk, G.M.: Study on metrological relations between instant tool displacements and surface roughness during precise ball end milling. Measurement **129**, 686–694 (2018)
11. Abellan-Nebot, J.V., Bruscas, G.M., Serrano, J., Vila, C.: Portability study of surface roughness models in milling. Procedia Manuf. **13**, 593–600 (2017)
12. Zhang, S.J., To, S., Wang, S.J., Zhu, Z.W.: A review of surface roughness generation in ultra-precision machining. Int. J. Mach. Tools Manuf **91**, 76–95 (2015)

13. Brammerz, P.H.: Die Entstehung der Oberflachenrauheit beim Feindrehen. Industrie Anzeiger **2**, 25–32 (1961)
14. Zong, W.J., Huang, Y.H., Zhang, Y.L., Sun, T.: Conservation law of surface roughness in single point diamond turning. Int. J. Mach. Tools Manuf **84**, 58–63 (2014)
15. Wojciechowski, S., Twardowski, P., Pelic, M., Barrans, S., Krolczyk, G.M.: Precision surface characterization for finish cylindrical milling with dynamic tool displacements model. Precis. Eng. **46**, 158–165 (2016)
16. Arizmendi, M., Fernandez, J., de Lacalle Lopez, L.N., Lamikiz, A., Gil, A., Sanchez, J.A., Campa, F.J., Veiga, F.: Model development for the prediction of surface topography generated by ball-end mills taking into account the tool parallel axis offset. Experimental validation. CIRP Ann. – Manuf. Technol. **57**, 101–104 (2008)
17. Zhanga, S., Zhoua, Y., Zhangb, H., Xionga, Z., To, S.: Advances in ultra-precision machining of micro-structured functional surfaces and their typical applications. Int. J. Mach. Tools Manuf **142**, 16–41 (2019)

Technological Aspects of Regeneration of Carbide End Mills

Paweł Twardowski[1]([⊠]), Patryk Welsant[2], and Marek Rybicki[1]

[1] Institute of Mechanical Technology, Poznan University of Technology, ul. Piotrowo 3, 60-965 Poznań, Poland
pawel.twardowski@put.poznan.pl

[2] Aesculap Chifa Sp. z o.o., ul. Tysiąclecia 14, 64-300 Nowy Tomyśl, Poland

Abstract. The article concerns selected technological aspects of carbide end mills regeneration. Geometry of cutting edges was measured (clearance angle α_0, rake angle γ_0 and cutting edge radius r_n) of commercial and regenerated mills. The measurements were taken for the initial cutting edge wear and after 30 min of machining. In addition, a visual evaluation of the cutting edge state of new and regenerated cutters before and after machining was made. There were no significant differences in geometry changes and wear of commercial and regenerated cutters after machining. An increase in the radius r_n and a decrease in the angle α_0 was observed after cutting. The cost of regeneration including grinding and re-coating of cutters was 17% of the price of a new tool with a diameter of 10 mm.

Keywords: Regeneration · Carbide end mill · Geometry · Wear · Cost of regeneration

1 Introduction

Nearly all the monolithic end mills used in production can be regeneration. In this process, you can keep almost all of their parameters (dimensions, geometry) or modify them depending on the client's needs, resulting e.g. from the necessity of using non-standard inserts in cutting tools or inserting additional pockets.

Professionally regenerated cutting tools are not inferior to the new ones. Their quality is almost identical, and this solution involves much lower costs than the purchase of a new counterpart [1, 5]. A well-made regeneration allows obtaining a tool life of at least 90% of the life of the new one. The cost of regenerated end mills made of cemented carbide varies in most cases within 15–50% of the new tool, depending on of the tool diameter mainly. The highest cost-effectiveness occurs in the regeneration of tools with large diameters due to high price of carbides reaching 80% of the tool price [2]. Taking into account five cycles of regeneration, end-user savings can reach 60–70% of tool costs. The best example of the regeneration process profitability is the automotive industry, where almost 100% of monolithic tools are regenerated. In case of repairing special tools, it is often a very important benefit for the customer to shorten the waiting time for the tool, because the regeneration time is shorter than making time of the new tool.

G. M. Królczyk et al. (Eds.): IMM 2019, LNME, pp. 55–63, 2020.
https://doi.org/10.1007/978-3-030-49910-5_6

Regeneration is not only a way to save money and reduce production costs. Thanks to this, the tools are fully used, and this also affects the protection of the environment. Due to regeneration of tools, a much lower amount of CO_2 is emitted than in the case of new production. The percentage data presented in this paragraph is the result of many years of cooperation with the industry of the author of this publication.

Regeneration of tools is becoming more and more popular. Many companies in Poland and around the world have specialized in this process. However, regenerated tools are often identified with a product of inferior quality, suitable only for roughing. Statistical dates on the steel processing market in Poland for the last 5 years shows that only 1,65% of enterprises are guided by the possibility of tools regeneration when choosing a their supplier (Fig. 1), and companies providing tool sharpening services is amounted to 19% (Fig. 2). The report [4] was based on surveys of about 500 Polish companies from group of small and medium-sized enterprises (up to 250 employees). These dates lead to conclusion that market for regenerated tools is at an early stage of development in Poland.

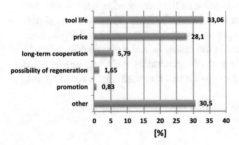
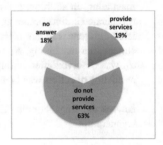

Fig. 1. The criterion for choosing a tool supplier [4].   **Fig. 2.** Percentage of companies providing tool sharpening services [4].

The article presents results related to regeneration of carbide cutters operating in industrial conditions. The data comes from a company from the medical industry located in Poland.

2 Aim, Range and Methodology of Research

The aim of the research was to compare macro, micro-geometry and wear of commercial and regenerated tools as well as to evaluate the cost-effectiveness of the regeneration process.

A fine-grained carbide end mill from Fraisa with a diameter of $D = 10$ mm and a catalog designation P5240450 (Fig. 3) and their regenerated equivalent with the assumed designation RE-P5240450 were used in the research.

Fig. 3. End mill ($z = 4$, $\lambda_s = 45°$, $\gamma_0 = 15°$).

The cutters were used for machining of heat-treated 55CrNiMoV7 steel, which is used on hot stamping dies and punches with a hardness of 350–400 HB. For fixing the end mills the REGO-Fix powRgrip holder system was used, which is a mechanical alternative for shrink-fit or hydraulic mounting. The workpiece was clamped with a vise, made of Lang, equipped with a quick assembly and disassembly system from the machine table. Milling by commercial and regenerated cutters was carried out with the same parameters recommended by the tool manufacturer:

- cutting speed $v_c = 75$ m/min,
- cutting depth $a_p = 15$ mm,
- cutting width $a_e = 1$ mm,
- feed per tooth $f_z = 0.035$ mm,
- feed speed $v_f = 335$ mm/min.

All cutters (commercial and regenerated) were working during $t_c = 30$ min, and then the cutters were subjected to geometric measurements.

The regeneration process of worn commercial milling cutters P5240450 was carried out on the Hawat's 2001 Eco numerical grinder. The regeneration process was carried out in such a way as to obtain the cutting edge geometry similar to or the same as for a commercial milling cutter before machining.

Measurements of macro, micro-geometry and wear of cutters were carried out on Alicona InfiniteFocus G5 microscope.

Figure 4 presents method of measuring of the end mills geometry. The measuring device is an optical system and thanks to the modern technology Real3D has the ability to measure all the features of the geometry of such cutting tools such as cutters and drills. It allows to automatically collect up to several dozen million measurement points in a very short time (1,7 million points per second) and create a 3D model on which measurements can be taken.

Five new and five regenerated end mills were used in geometry research. The research contained analysis of rake angle, clearance angle and cutting edge radius. The results of geometry measurement of the commercial cutter before machining were taken as the reference values on the basis of which further analyzes were made.

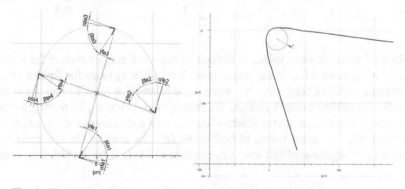

Fig. 4. The method of the α_0 and γ_o angles as well as the r_n radius measurement.

In order to evaluate wear of the end mills after machining, the mills were dismounted from the holder and subjected to optical analysis under a microscope of 20-magnification approximately.

3   Analysis of Test Results

3.1   Geometry of End Mills

Table 1 shows results of the geometry measurement of the commercial end mill before machining. As can be seen, measurements results of cutting edge radius r_n, clearance angle α_0 and rake angle γ_0 of the commercial end mill before machining are very similar for each of the tooth.

All remanufactured end mills had the same geometry as new end mills. Results of the geometry after 30 min of cutting for the first examined commercial cutter are shown in Table 2 and for the regenerated cutter in Table 3. The influence of the machining process on the cutters geometry is noticeable. The cutting edge radius r_n increased several times and the α_0 angle decreased its value, both for commercial and regenerated cutter. The largest geometry changes, relative to commercial end mill before machining, were observed for the rake angle γ_0 of regenerated cutter.

Table 1.   Geometry of a commercial end mill before cutting.

Tooth number	Cutting edge radius r_n [μm]	Clearance angle α_0 [°]	Rake angle γ_0 [°]
1	3,1	8,6	19,4
2	2,9	8,5	19,1
3	3,3	8,7	19,8
4	3,1	8,5	19,3
x_{avg}:	3,1	8,6	19,4
σ:	0,2	0,1	0,3
min.:	2,9	8,5	19,1
max.:	3,3	8,7	19,8

Figure 5 presents mean values of the cutting edge radius r_n of each of the analyzed cutters. The horizontal line in the graph shows level of the r_n value for the new ($VB_B = 0$ mm) commercial milling cutter, the mean value of which is $r_n = 3,1$ μm. As noted in the study, the milling process changes the cutting edge radius r_n, increasing its value. When comparing measurements results of commercial and regenerated end mills, it can be noticed that the radius values in both cases are very similar. On this basis it was concluded that regenerated and new cutters wears similarly from the radius r_n point of view.

Table 2. Geometry of the commercial end mill No. 1 after cutting time $t_c = 30$ min.

Tooth number	Cutting edge radius r_n [μm]	Clearance angle α_0 [°]	Rake angle γ_0 [°]
1	15,5	4,8	19,7
2	10,1	1,3	25,6
3	8,4	6,3	18,3
4	14,6	6,8	15,9

Table 3. Geometry of regenerated end mill No. 1r after cutting time $t_c = 30$ min.

Tooth number	Cutting edge radius r_n [μm]	Clearance angle α_0 [°]	Rake angle γ_0 [°]
1	11,7	5,6	10,7
2	10,8	5,5	9,9
3	13,8	5,9	11,3
4	10,2	5,7	11,2

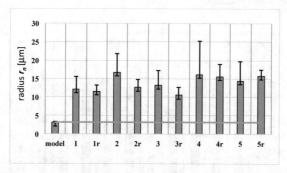

Fig. 5. Average values of radius r_n for all analyzed cutters (1, 2, … number of new end mills after 30 min of cutting, 1r, 2r, … number of regenerated end mills after 30 min of cutting)

Analyzing the measurements results of the clearance angle α_0, changes of its value are seen after machining relative to reference milling cutter (Fig. 6). The clearance angle mean value is $\alpha_0 = 8,6°$ before machining. The α_0 angle for commercial and regenerated cutters after machining varies within a few degrees, below the value for the reference cutter. The fact that the clearance angle decreases with increasing of tool wear seems natural because the wear process (wear on the flank face) causes the clearance angles to reach locally zero values.

In case of rake angles γ_0, a significant difference in values for commercial and regenerated cutters was observed (Fig. 7). Value of the angle for reference cutter was $\gamma_0 = 19,4°$. After cutting time $t_c = 30$ min, the rake angles of the commercial cutters changed slightly by approximately 2°. In case of regenerated cutters, the rake angles decreased by more than 15°. The probable cause can be the fact that during regeneration

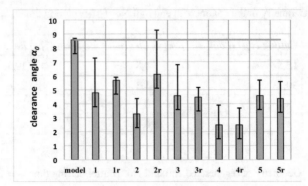

Fig. 6. Average values of clearance angles α_0 for all analyzed cutters (1, 2, … number of new end mills after 30 min of cutting, 1r, 2r, … number of regenerated end mills after 30 min of cutting).

of the cutters grinding process was poorly carried out, giving the cutters lower values of the rake angles. It is hard to find another reason for such large differences in the γ_0 values, knowing earlier that the clearance angles α_0 for commercial and regenerated cutters behave as intended. The value of the grinding wheel angle should be verified during the regeneration process.

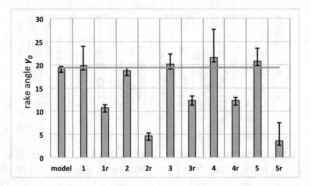

Fig. 7. Average values of rake angles γ_0 for all analyzed cutters (1, 2, …number of new end mills after 30 min of cutting, 1r, 2r, … number of regenerated end mills after 30 min of cutting).

3.2 Wear of End Mills

Table 4 shows images of edges of three exemplary commercial and regenerated end mills after cutting time $t_c = 30$ min. As can be seen, commercial and regenerated cutters are characterized by similar wear on the main cutting edge. There were also no significant differences between the work of commercial and regenerated cutters.

Table 5 presents model of each main cutting edge generated by the optical device. When analyzing all cutting edges of commercial and regenerated end mills, it should be noted that there are no significant differences both from the point of view of radius r_n

Table 4. Wear of exemplary end mills after cutting.

	end mill 1	end mill 3	end mill 5
Commercial mill			
	end mill 1r	end mill 3r	end mill 5r
Regenerated mill			

and wear on flank as well as rake faces. Similar relations were observed for other milling cutters by analyzing their images after the cutting time $t_c = 30$ min. Therefore, it can be concluded that there is no tool life difference between commercial and regenerated end mills. End mills working in the same conditions, at the same time, achieve similar results

3.3 Cost of Regeneration

The end mills regeneration process is much cheaper compared to the purchase of new end mill, which has a positive effect on the reduction of tooling costs in production plants. The cost of regeneration with coating was 44 PLN, which is 17% of the price of the new end mill (Fig. 8).

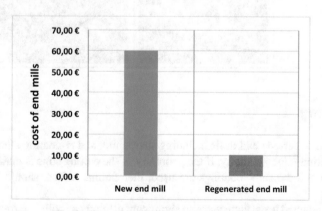

Fig. 8. Cost of new and regenerated end mills.

Table 5. Wear of individual cutting edges of end mill 1.

Tooth number	Commercial mill	Regenerated mill
1		
2		
3		
4		

4 Summary

Regeneration of carbide end mills including sharpening and re-coating allows to obtain the same cutting-edge geometry. If the geometry of the cutting edge is maintained after regeneration and the same coatings are used, then similar effects should be expected after milling.

In the conducted tests, there were no significant differences both in terms of radius r_n and abrasive wear on flank and rake faces after milling with commercial and regenerated end mills.

It was found that during machining sharp cutting edges of the cutters ($r_n = 3,1$ μm) increase their radius several times and the clearance angles decrease. This phenomenon is related to the abrasive wear of the tool, mainly on the flank face and within the cutting edge.

It is possible to monitor the condition of the cutting tools in real time, but these methods are related to another issue and do not relate to the theme of this article [7, 8].

References

1. Chen, J.Y., Lee, B.Y., Chen, C.H.: Planning an analysis of grinding processes for end mills of cemented tungsten carbide. J. Mater. Process. Technol. **201**, 618–622 (2008)
2. Denkena, B., Dittrich, M.A., Heuwold, N., Liu, Y., Theuer, M.: Automatic regeneration of cemented carbide tools for a resource efficient tool production. Procedia Manuf. **21**, 259–265 (2018)
3. Górski, E.: Poradnik narzędziowca, Warszawa (PWN) (2017)
4. Raport: Rynek obróbki stali w Polsce. Elamed Media Group (2019)
5. Uhlmann, E., Hubert, C.: Tool grinding of end mill cutting tools made from high performance ceramics and cemented carbides. CIRP Ann. Manuf. Technol. **60**(1), 359–362 (2011)
6. Wronska, I., Czechowski, K., Toboła, D., Bednarski, P., Królicka, B.: Możliwości regeneracji narzędzi skrawających powlekanych. Mater. Eng. **4**(182), 676 (2011)
7. Bustillo, A., Rodriguez, J.J.: Online breakage detection of multitooth tools using classifier ensembles for imbalanced data. Int. J. Syst. Sci. **45**(12), 2590–2602 (2014)
8. Mikołajczyk, T., Nowicki, K., Bustillo, A., Pimenov, D.Yu.: Predicting tool life in turning operations using neural networks and image processing. Mech. Syst. Signal Process. **104**, 503–513 (2018)

Influence of Milling Parameters on Tribological Properties of Aluminum-Ceramic Composite

Dariusz Korzeniewski$^{(\boxtimes)}$ and Natalia Znojkiewicz

Faculty of Mechanical Engineering and Management, Poznan University of Technology,
Poznań, Poland
dariusz.korzeniewski@doctorate.put.poznan.pl

Abstract. This article presents the results of tribological tests carried out for the pair: aluminum-ceramic composite type F3S.10S - stainless steel. Stand-up tribological tests were carried out under dry friction conditions on a *pin-on-disc* tribometer. 3D roughness tests for Sa and Sz data were performed both before and after the tribological examination on the tribometer. These results describe how machining parameters of type v_c and f affect the roughness of the machined surface roughness. On the basis of the conducted tests, it was found that the feed has the greatest impact on the change of both the coefficient of friction μ and the friction force F_t. It was also shown that v_c cutting speed has a non-monotonic effect on the parameters of Sa and Sz roughness of milled samples. With an increase in feed f, both for Sa and Sz parameters, an upward trend was observed. From the tribological point of view, it was noticed that after the tribometer test, the surface roughness increased several times. Also for the weight loss test Δm of the tribological pair tested, the input parameters $v_c = 700$ m/min and $f = 0.20$ mm/rev cause the smallest weight loss of the composite kinematic composite - stainless steel, which is the optimal combination of milling parameters from the tribological point of view.

Keywords: Coefficient of friction · Metal matrix composite · Tribometer · Surface roughness · Weight loss

1 Introduction

Metal matrix composites (MMC) as new generation materials; they prove very profitable in various industries, such as biomedicine and the aerospace industry. To obtain a valuable modification of various material properties, the metal matrices are reinforced with additional phases based on the chemical and physical properties required in the operating conditions. The presence of reinforcements in MMC improves the physical, mechanical and thermal properties of the composite; however, this results in significant machining problems, such as high tool wear and insufficient surface roughness finish. The interaction between the tool and the hard matrix-reinforcing particles induces complex deformation behavior in the MMC structure [1].

© Springer Nature Switzerland AG 2020
G. M. Królczyk et al. (Eds.): IMM 2019, LNME, pp. 64–74, 2020.
https://doi.org/10.1007/978-3-030-49910-5_7

Materials based on light metal alloys are a group of materials most forward-looking among construction materials. Their advantage is the low weight of products made of them, while maintaining the required mechanical properties, and therefore the safety of exploitation. Composite materials with a metal matrix can be reinforced with particles, dispersion particles or fibers. However, the greatest interest is in the metal-particle composite materials due to good mechanical properties as well as good abrasion resistance [2].

The development of aviation and space technologies forced the necessity to develop new, materials resistant to elevated temperatures that meet the requirements for high strength and low density [3].

Aluminum ceramic composites with improved mechanical and chemical properties are necessary and needed in the aerospace and automotive industries. The composite with aluminum matrix reinforced with alumina (Al_2O_3) has good tribological properties. In the work of Kanthavel et al. [4] attempted to develop a new material using the powder metallurgy technique by adding molybdenum disulphide (MoS_2), which acts as a solid lubricant. Microstructures, material combinations, wear and friction properties were analyzed using scanning electron microscopy, EDX and a pin-on-disc wear tester. The research revealed that the appropriate selection of the input parameters results in a minimum wear and friction coefficient μ.

In the work of Hasegawa et al. [5] investigated the tool wear and cutting resistance in which they showed that the tool wear and cutting resistance were greater in the case of an alloy reinforced with harder ceramic particles for alloys containing Al_2O_3 or SiC. Tool wear was mainly caused by mechanical abrasion caused by ceramic particles in the feet. The main cutting force resulted from the frictional force caused by the tool wear. However, the consumption of sintered diamond tools was small, even for alloys containing Al_2O_3 or SiC, suggesting that tool wear and cutting resistance were also reduced if the hardness of the particles in the tool exceeded the hardness of the work material.

Przestacki et al. [6] presented studies on the formation of a surface roughness layer in A359/20SiCP (MMC) metal-ceramic composites during laser assisted turning, including the deposition of SiC particles in a liquid matrix. The developed model included the effect of gravity, displacement, resistance of the liquid matrix and the centrifugal force of the rotating working material. Based on the proposed approach, the instantaneous deposition rate as well as the depth of the deposited SiC particles were calculated. As a result, the applied model enabled the selection of the effective depth of cut and angular distance of the tool from the laser beam, which can improve the quality of the work surface roughness and the performance properties of the composite.

Paper [7] investigated the effect of finishing on the tribological properties of plasma MMC composite coatings. The use of Ni-5% Al-15% Al_2O_3 sprayed plasma coatings has been proposed. Shell surfaces roughness were formed by turning and polishing. The tribological properties of composite materials depended on the proportion and size of the reinforcing phase particles contained in the metal matrix. A reinforced composite with very small dimensions and quantities can lead to increased wear and increase in the coefficient of friction μ.

Work [8] presented the development and application of a tribological test rig mimicking typical contact geometries and high impact loads associated with infeed rotary swaging for process-independent tribological investigations. The development comprises the designing process based on rotary swaging process simulations, setup assembly, and calibration procedure.

Paper [9] investigated the design and manufacturing of microstructured surfaces for improved wear resistance of forming tools and reduced friction in powertrain components. The potential of microstructures in the improvement of adhesion strength and resistance to delamination of hard coatings under the severe thermomechanical service conditions of hot forging tool surfaces is discussed.

In the paper [10] Liu et al. focused on mechanical properties, fracture mechanism and machinability of ceramic-reinforced MMCs, with significant emphasis on the chip formation mechanism considering different dominant effects, such as materials strengthening mechanisms, micro-structural effect, size effect and minimum chip thickness effect.

Literature review shows that despite the conducted research on tribological properties and machinability of MMC composites, there have been few researches regarding the influence of milling conditions of the aluminum-ceramic composite on its tribological properties. Therefore, in this work a number of studies on the impact of milling parameters of the aluminum-ceramic composite F3S.10S on the tribological conditions of cooperation for a composite kinematic composite - stainless steel were made. In the conducted tests, a stainless steel disk was used to assess the tribological conditions in the sample system - counter-sample during friction.

2 Experimental Details

2.1 Research Method

The work material was an aluminum-ceramic composite type F3S.10S with a matrix of aluminum alloy reinforced with SiC particles. The most useful features of these composites are their high strength, stiffness and abrasion resistance.

The tests were carried out on a three axis milling center with a maximum spindle speed $n = 30.000$ rev/min.

As a cutting tool, a monolithic endmill has been used with a diamond cover, ø $10/10 \times 72/22$ R0.50, inclination angle of the main cutting edge $\lambda = 40°$, angle of attack $\gamma = 15°$, number of fresh cutter blades with diamond coverage $t = 3$, Cooling – YES, milling cutter fixed in a heat-shrinkable sleeve (Fig. 1).

The milling parameters for samples no. 1–16 of the aluminum-ceramic composite F3S.10S are shown in Table 1.

Fig. 1. Solid carbide milling cutter with diamond coverage

Table 1. Parameters of milling samples No. 1–16 of the aluminum-ceramic composite F3S.10S

No.	v_c [m/min]	f [mm/rev]	n [rev/min]	v_f [mm/min]	No. of blades	a_p [mm]	a_e [mm]	d [mm]
1	100	0.01	3185	32	3	10	0.1	10
2	100	0.05	3185	159	3	10	0.1	10
3	100	0.10	3185	318	3	10	0.1	10
4	100	0.20	3185	637	3	10	0.1	10
5	300	0.01	9554	96	3	10	0.1	10
6	300	0.05	9554	478	3	10	0.1	10
7	300	0.10	9554	955	3	10	0.1	10
8	300	0.20	9554	1911	3	10	0.1	10
9	700	0.01	22293	223	3	10	0.1	10
10	700	0.05	22293	1115	3	10	0.1	10
11	700	0.10	22293	2229	3	10	0.1	10
12	700	0.20	22293	4459	3	10	0.1	10
13	900	0.01	28662	287	3	10	0.1	10
14	900	0.05	28662	1433	3	10	0.1	10
15	900	0.10	28662	2866	3	10	0.1	10
16	900	0.20	28662	5732	3	10	0.1	10

Tribological experimental research was carried out on the Tribotester *pin-on-disc* tribometer, which enables tribological pair cooperation in the sample - counter-sample system. The construction of the test bench allows the measurement of the friction coefficient μ under the effect of the normal force F_n at time T (in the range of input quantities). During the friction tests, the average values of friction coefficient μ and friction force F_t were determined based on tribological measurements.

3D roughness tests for Sa (Sa - arithmetic average of the ordinates of the 3D roughness profile) and Sz (Sz – average maximum height of the 3D roughness profile) parameters were performed on the profilograph.

2.2 Materials of Tribological Pair

The friction tests were carried out in a system that is a classical tribological pair - a sample - a counter-sample. In the stand tribological tests, as samples from 1 to 16, an aluminum-ceramic composite type F3S.10S with dimensions of $10 \times 10 \times 20$ mm. The anti-test disc was made as a type 0H17N12M2 stainless steel disc with a flat surface roughness. Such a surface roughness configuration allowed the unit to obtain a unit pressure of $F_N = 10$ N at the friction stage, comparable to those encountered in the friction process of two materials.

Its physical properties are shown in Table 2 and its chemical composition is shown in Table 3 [11].

Table 2. Physical properties of DURALCAN^{TM} F3S.10S composite:

Physical properties of DURALCAN^{TM} F3S.10S composite		
Density	[g/cm^3]	2.71
Melting temperature	[K]	1003
Highest tensile strength	[MPa]	338 (310)
Optimal tensile strength	[MPa]	303 (283)
Elongation	[%]	1,2
Modulus of elasticity	[GPa]	86.2
Rockwell hardness	[HRB]	73

Table 3. The chemical composition of the DURALCAN^{TM} F3S.10S matrix alloy:

Chemical composition DURALCAN^{TM} type F3S.10S									
	Si	Fe	Cu	Mn	Mg	Ni	Ti	Others	Al
Alloy F3S	8.50–9.50	0.20 max	0.20 max	–	0.45–0.65	–	0.20 max	0.03 max 0.10 together	Rest

The stainless steel material is a type 0H17N12M2 disk with dimensions of 70 × 6 mm with a central hole needed for mounting on the tribometer.

2.3 Scope of Tests

The speed of the disk $n = 120$ rev/min, Pressure force $F = 10$ N (Dry friction), Test time $T = 3600$ s,

3D roughness tests were carried out on the profilograph:

The length of the measuring section $Lt = 4.80$ mm. The length of the elementary segment 0.8 mm, Area of measurement 3D, $X = 4.8$ mm, $Y = 2.5$ mm.

3 Results and Discussion

In the first stage of the research, the coefficients of friction μ and the friction force F_t in function (v_c, f) were analyzed, which are shown in Fig. 2.

Based on the conducted tests, it was found that the lowest coefficient of friction μ was 0.051 with milling parameters $v_c = 700$ m/min and $f = 0.10$ mm/rev for sample No. 11.

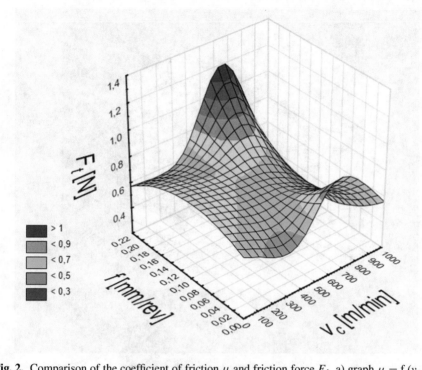

Fig. 2. Comparison of the coefficient of friction μ and friction force F_t, a) graph $\mu = f(v_c, f)$, b) graph $F_t = f(v_c, f)$

The highest friction coefficient μ was 0.151 with milling parameters $v_c = 700$ m/min. and $f = 0.20$ mm/rev for sample No. 12. The results for the friction force F_t are similar. The smallest friction force F_t was 0.468 N with milling parameters $v_c = 700$ m/min and $f = 0.10$ mm/rev for sample No. 11. The largest friction force F_t was 1.250 N with milling parameters $v_c = 700$ m/min and $f = 0.20$ mm/rev for sample No. 12.

The cutting speed in both cases was $v_c = 700$ m/min and had an indirect effect on the average coefficient of friction μ and friction force F_t. In the v_c range from 100 m/min to 900 m/min, the cutting speed does not have a significant influence on the coefficient of friction μ and friction force F_t, but for about 700 m/min the highest values of coefficients of friction μ and friction force F_t were obtained. On the other hand, a very large influence on both these values of μ and F_t had a feed. The highest values of μ and F_t were obtained for sample No. 11 milled with $f_{11} = 0.10$ mm/rev and for sample No. 12 it was $f_{12} = 0.20$ mm/rev.

Analyzing the presented values, it can be concluded that when the feedrate changes from 0.10 mm/rev to 0.20 mm/rev, both the friction coefficient μ and the friction force F_t increase about 3 times. Thus, the feed has the greatest influence on the friction coefficient μ and friction force F_t.

In the next stage of the research, the comparison of 3D roughness values for Sa and Sz in the function of $f(v_c, f)$ before the test on the tribometer and after the tribometer was analyzed, which are shown in Figs. 3 and 4.

a)

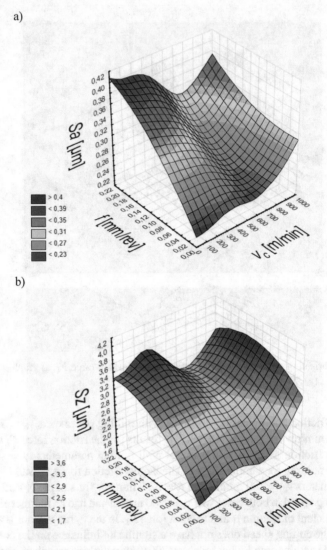

b)

Fig. 3. Comparison of roughness values before testing on the tribometer, a) for $Sa = f(v_c, f)$, b) for $Sz = f(v_c, f)$

The smallest roughness Sa and Sz was obtained after milling at the cutting speed $v_c = 700$ m/min, in the entire tested feed range. The influence of the cutting speed v_c on the values of roughness parameters Sa and Sz for samples 1–16 is non-monotonic. With an increase in feed f, both for the values of Sa and Sz, an upward trend was observed. However, it was also noticed that at the cutting speed of $v_c = 700$ m/min, we have the smallest value of the Sa and Sz roughness parameters.

a)

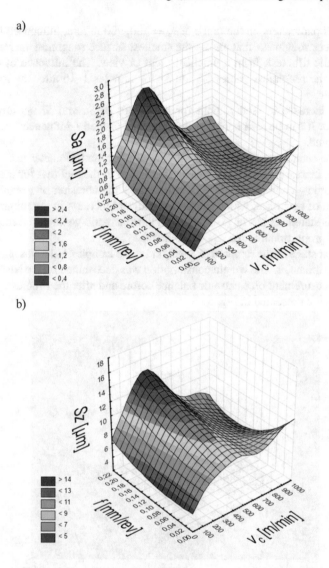

b)

Fig. 4. Comparison of roughness values after testing on a tribometer, a) for $Sa = f(v_c, f)$, b) for $Sz = f(v_c, f)$

Assessment of surface roughness of composite samples after testing on a tribometer showed that the largest increases in the roughness parameter Sa were obtained for sample No. 8 (multiplicity × 7.16, at $v_c = 300$ m/min and $f = 0{,}20$ mm/rev), and the parameter Sz for sample No. 10 (fold × 4.34, at $v_c = 700$ m/min and $f = 0.05$ mm/rev). On the other hand, the smallest increments of the surface roughness parameters were observed for sample No. 15 (multiplicity × 2.29, at $v_c = 900$ m/min and $f = 0.10$ mm/rev), type Sz for sample No. 15 (multiples × 1.64; at $v_c = 900$ m/min and $f = 0.10$ mm/rev).

The input parameters, in the range of $v_c = 700$–900 m/min, influenced the generation of surfaces roughness that ensure the smallest surface roughness increments after performing the tribotest, from a practical point of view. The influence of the cutting speed v_c on the roughness values Sa and Sz for samples 1–16 after the tribotester is nonmonotonic.

With an increase in feed f, both for the values of Sa and Sz, an upward trend was observed. It can be said here that the influence of feed f influences indirectly the roughness value.

As can be seen from figures no. 3 and no. 4, after the tribometer test, the Sa and Sz roughness increases are several times. This is due to the fact that for the kinematic pair, the sample - counter-sample was followed by the abrasion of irregularities and the formation of new ones, which caused a significant increase in the roughness of the samples. This statement has its justification in the kinematic-geometric formula that the roughness is proportional to the square of the feed f.

In the next stage of the research, the weight loss of samples 1–16 was analyzed after testing on a tribometer. The weight consumption was determined on the basis of weight difference measurement on electronic balance before and after the tribotest. The results of these tests are presented in Fig. 5.

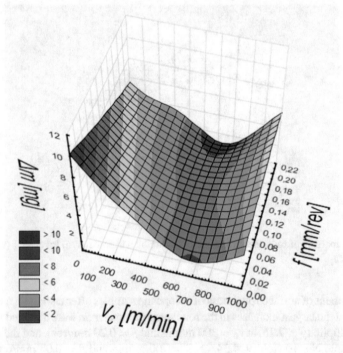

Fig. 5. Mass loss Δm of samples 1–16, spatial diagram $\Delta m = f(v_c, f)$

After the tests carried out on the tribometer, it was found that the largest loss in mass occurs for sample No. 2, which is $\Delta m_2 = 9.1$ mg. The input parameters that were $v_{c2} = 100$ m/min and $f_2 = 0.05$ mm/rev were influenced by the input parameters for generating the surface roughness that provided the largest mass loss after the tribotest. However, the smallest mass loss occurs for sample No. 12 and is $\Delta m_{12} = 1.1$ mg. The input parameters, which were for $v_{c12} = 700$ m/min and $f_{12} = 0.20$ mm/rev, influenced the generation of the surface roughness providing the smallest weight loss after the tribotest. It follows from this conclusion that the optimal selection of input parameters such as cutting speed v_c and feed f can improve the tribological conditions of cooperation for the composite - stainless steel pair.

4 Summary

Based on the experimental studies carried out, the following conclusions were made:

- After testing on the tribometer, the surface roughness increases several times. Both the cutting speed v_c and the feed f affect the surface roughness Sa and Sz. In the case of feed increase f, the roughness growth trend is observed. And in the case of increasing the cutting speed v_c this effect is non-monotonic.
- The tests have shown the connection of milling parameters with the surface roughness after the tribing test and may also affect the weight loss of the tribological sample-counter-sample pair. Therefore, the optimal selection of milling input parameters may affect the favorable tribological properties of the composite-stainless steel pair.
- The input parameters type $v_c = 700$ m/min and $f = 0.20$ mm/rev cause the smallest weight loss of the samples used for the tribological conditions of cooperation for the composite - stainless steel composite pair.

References

1. Aramesh, M., Shaban, Y., Balazinski, M., Attia, H., Kishawy, H.A., Yacout, S.: Survival life analysis of the cutting tools during turning titanium metal matrix composites (Ti-MMCs). In: 6th CIRP International Conference on High Performance Cutting, HPC2014, Procedia CIRP, vol. 14, pp. 605–609 (2014)
2. Dobrzański, L.A., Włodarczyk, A., Adamiak, M.: Materiały kompozytowe o osnowie stopu aluminium AlCu4Mg1 wzmacniane cząstkami ceramicznymi. In: 12th International Scientific Conference Achievements in Mechanical & Materials Engineering, AMME (2003)
3. Grzesik, W.: Podstawy skrawania materiałów konstrukcyjnych, Wydanie 3 zmienione i uaktualnione, Wydawnictwo Naukowe PWN SA Warszawa 2017, str. 311–314 (2018)
4. Kanthavel, K., Sumesh, K.R., Saravanakumar, P.: Study of tribological properties on Al/Al$_2$O$_3$/MoS$_2$ hybrid composite processed by powder metallurgy. Alex. Eng. J. **55**, 13–17 (2016)
5. Hasegawa, T., Nishiwaki, N., Zhang, L.-P., Miura, T.: Wear of carbide tool in turning of mechanically alloyed aluminum-ceramic particle composite materials. J. Jpn. Inst. Light Met. **44**(6), 359–364 (1994)

6. Przestacki, D., Szymanski, P., Wojciechowski, S.: Formation of surface layer in metal matrix composite A359/20SiCP during laser assisted turning. Compos.: Part A **91**, 370–379 (2016)
7. Charchalis, A., Starosta, R.: The influence of finishing on the tribological properties of plasma sprayed MMC coatings. J. KONES Powertrain Transp. **23**(1) (2016)
8. Böhmermann, F., Riemer, O.: Development and application of a test rig for tribological investigations under impact loads. CIRP J. Manuf. Sci. Technol. **19**, 129–137 (2017)
9. Schubert, A., Neugebauer, R., Sylla, D., Avila, M., Hackert, M.: Manufacturing of surface microstructures for improved tribological efficiency of powertrain components and forming tools. CIRP J. Manuf. Sci. Technol. **4**(2), 200–207 (2011)
10. Liu, J., Li, J., Xu, C.: Interaction of the cutting tools and the ceramic-reinforced metal matrix composites during micro-machining: a review. CIRP J. Manuf. Sci. Technol. **7**(2), 55–70 (2014)
11. Duralcan MMC Product Data Sheet. Accessed 10 May 2016

Analysis of the Turning Process to Optimize Energy Consumption

Agnieszka Terelak-Tymczyna[✉], Karol Miądlicki, and Marcin Jasiewicz

Instytut Technologii Mechanicznej, Zachodniopomorski Uniwersytet Technologiczny
w Szczecinie, al. Piastów 19, 70-310 Szczecin, Poland
Agnieszka.Terelak@zut.edu.pl

Abstract. Electricity is one of the most widely used forms of energy in companies, especially in machining. Nowadays, machine tools are equipped with the latest technical solutions in the field of drives and control. Braking energy recovery systems and software for evaluating energy consumption are widely used in drive controllers. The paper presents an analysis of the consumption of both active and reactive electric energy in the turning process for different spindle speeds and different depths of the machined layer. The aim of this article is to present the results of the analysis of electric energy consumption divided into active and reactive energy during the turning process with various machining parameters such as spindle speed and depth of cut. According to obtained data it is better machining with high speed and low depth of cut.

Keywords: Machining · Turning · Energy efficiency · Energy management

1 Introduction

Electricity is one of the basic forms of energy used in today's metal industry, particularly in the machining processes. According to the literature [1] the largest share in the energy consumption of the machine tool have the cooling and lubrication system (31%) and the hydraulic system (27%) as presented in Fig. 1.

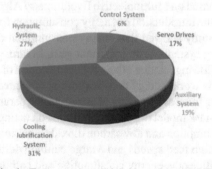

Fig. 1. Energy consumption in machine tool [1].

© Springer Nature Switzerland AG 2020
G. M. Królczyk et al. (Eds.): IMM 2019, LNME, pp. 75–84, 2020.
https://doi.org/10.1007/978-3-030-49910-5_8

The issue of energy consumption in the machining process has been the subject of many studies. Most of them consider energy consumption in the context of machining parameters, planning machining process, planning including artificial intelligence methods and modeling of the machining process [2].

The impact of cutting parameters on the energy efficiency of the machining process was examined considering, among others type of materials processed, machine tools used, optimization possibilities and indicators of sustainable development.

Improving the energy efficiency of machining process by appropriate selection of process parameters for machining materials in the form of mild steel, aluminum and brass dealt Anand et al. [3]. The study on the improvement of energy efficiency also dealt Chudy, who identify a strategy for saving energy for the rolling of ductile iron examining the effect of cutting the energy capacity [4] and strategies for the treatment sequence with respect to the required roughness of the surface after thorough turning and burnishing of hardened steel [5]. Changes in energy consumption at different cutting parameters of hardened steel are also presented in [6]. Whereas in [7], changes in the value of cutting force and process energy after turning steel hardened with blades with different knife radius were analysed.

Reduction of electricity consumption was also considered when using various machine tools. Diaz et al. investigate reduction of energy consumption during final milling in three aspects: process parameters, kinetic energy recovery system and increase of efficiency through sensor integration with the monitoring scheme, was considered in [8]. In research, authors noted that the energy consumed by the machine tool is essentially independent of load and depends on the time of the process [8]. However, from research conducted by Mori et al. [9] shows that changing cutting conditions reduces energy consumption. Tests were carried out for drilling, face-end milling and deep hole machining.

A lot of scientific research has also been devoted to the possibilities of optimizing the machining process using modern methods [10], including artificial intelligence algorithms for planning. Such research was carried out by, among others Al Hazza et al., Which in publication [11] authors presents the optimal parameters of continuous processing in relation to the minimum cost of energy consumption while maintaining surface roughness in the range of acceptance using a multi-task genetic algorithm. Genetic algorithms were also used by Datta et al. in [12], where he presented optimization of machining parameters based on Multiobjective Evolutionary Algorithm. Newmann et al. Dealt with research on the introduction of energy consumption for planning the machining process [13]. The study showed that energy consumption can be added to multi-criteria process planning systems as an important goal. In turn, process planning using an adaptive approach allowing, among others for optimization of energy consumption is presented in [14]. In this work, energy consumption is estimated analytically based on the instantaneous cutting force as a function of actual cutting parameters. Whereas Zhang et al. developed a multi-task model of process optimization taking into account high efficiency, low energy consumption and low carbon dioxide emission was presented in [15]. The articles show that high feed speeds and a large cutting width can benefit all parameters considered, i.e. efficiency, energy consumption and carbon dioxide emissions, if the limits can be met.

In the literature we will also find online optimization methods using real-time control. Research on a component-based energy modelling methodology to implement the online optimization needed for real-time control can be found in the work Shin et al. [2].

In recent years, there have also been publications on sustainable development and related scientific research having regard to environmental indicators in the field of energy efficiency in metal shaping processes. This issue can be found in the work of Ingarao [16], in which he conducted a quantitative analysis, in relation to the level of the unit process, improvement of environmental indicators at different process parameters, machine architecture, the process used and the production process itself. The indicator analysis was also carried out by the team at the Institute of Mechanical Technology in Szczecin, which in [17] proposed energy indicators for the roller turning process.

In the analysis of electricity consumption by a machine or any other device, not only the part of energy that is used for work, i.e. active energy, should be considered. Reactive energy, which is often negligible in this type of analysis, is equally important from an economical and energy point of view [18].

Modern machine tools are equipped with a whole range of solutions for electric drives and control. The drives commonly use energy recovery systems during braking, as well as additional systems and software that allow the current assessment of electricity use. However, it is only recently (2013) that machine tool manufacturers have proposed additional systems for reactive energy consumption and reactive power compensation systems.

The aim of this article is to present the results of the analysis of electric energy consumption divided into active and reactive energy during the turning process with various machining parameters such as spindle speed and depth of cut.

For this purpose, 8 different settings of the machining process presented in Table 1 were analyzed. The following parameters were considered: f - feed rate, n - spindle speed, ap - depth of cut.

Table 1. Selected machining parameters

f, mm/rev	n, rpm	ap, mm
0.1	1200	0.25
	1400	0.25
	1600	0.25
	1200	0.5
	1400	0.5
	1600	0.5
	1200	1
	1400	1
	1600	1

2 Subject of Study

The research was conducted on a TAE-35N "Hanka" lathe Nominal data of the machine tool drives are presented in Table 2.

Table 2. Nominal data for TAE-35N machine drives.

Drive	Power, kW	Speed, rpm	Torque, Nm
axis X1	2.5	3000	12
axis X2	1.0	4000	4
axis Z1	2.5	3000	12
Spindle	11	1500–6000	

For a correct analysis of the energy consumed by machine tool, the nature of the load connected to the AC mains must be considered. In machine tool drives, the current is phase shifted in relation to voltage, which results in apparent energy consumption (active and reactive).

2.1 Research Stand

Researches of electricity consumption by the machine tool were carried out using the Lumel ND 20 recorder. Current measurement was performed using current transformers, while voltage was measured directly. The data was recorded in 1-s time interval. The connection diagram of Lumel ND 20 recorder is shown in Fig. 2 and the analyzed workpiece is shown in Fig. 3.

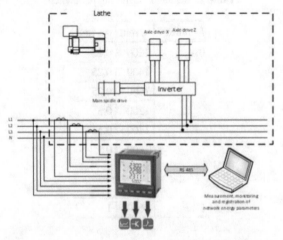

Fig. 2. Connection diagram of Lumel ND 20 recorder

Fig. 3. Analyzed workpiece machined with different parameters

2.2 Measurement Results

The results of experimental measurements of electric energy consumption for selected cutting parameters are presented in Figs. 4, 5, 6, 7, 8, 9, 10 and 11.

Analyzing the use of active P, reactive Q and apparent power S for machining using different rotational speeds, it can be noted that:

- for depth of cut. $ap = 0.25$ mm, the highest power values were recorded for machining at spindle speed of 1600 rpm, where the consumption of apparent power is in the range 3000–4200 VA, active power in the 600–1000 W and similarly reactive power in the range 600–1000 VAR (Fig. 4)
- for depth of cut. $ap = 0.5$ mm, the highest power consumption was observed when machining at a spindle speed of 1600 rpm. Apparent power takes values in the range of 3000–7500 VA, active power is used in the 800–1200 W range, reactive power in the 800–2000 VAR range. It should be noted that for the spindle speed at 1400 and 1600 rpm in the first 2–3 s of machining, there is a higher consumption of reactive power than the active one (Fig. 5)

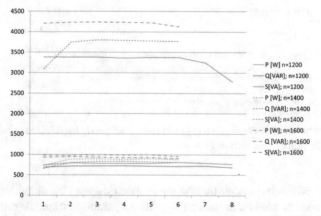

Fig. 4. Consumption of active P, reactive Q and apparent S power for selected spindle speeds and constant depth of cut ap $= 0.25$ mm and $f = 0.1$

– for depth of cut. $ap = 1$ mm, the highest power values were observed when machining at a spindle speed of 1600 rpm, however for the spindle speed at 1200 rpm, a significant increase in apparent power demand can be observed at the end of machining. Apparent power has a value in the range of 3500–7500 VA, active power is consumed in the 800–1000 W range, reactive power in the 800–2000 VAR range. It should be noted that for the spindle speed of 1200 rpm in the 7 s machining time, the reactive power consumption is lower than the active one, which may indicate the spindle drive braking (Fig. 6).

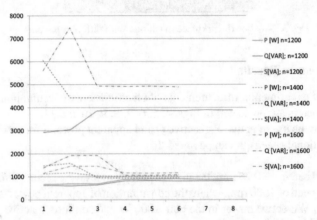

Fig. 5. Consumption of active P, reactive Q and apparent S power for selected spindle speeds and constant depth of cut ap $= 0.5$ mm and $f = 0.1$

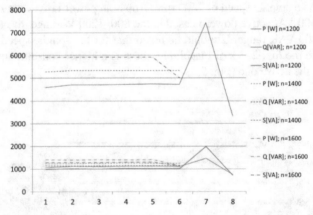

Fig. 6. Consumption of active P, reactive Q and apparent S power for selected spindle speeds and constant depth of cut $ap = 1$ mm and $f = 0.1$.

The obtained results indicate that the lowest electric energy consumption, both active and passive, was obtained for the machining with the parameters: spindle rotational speed 1400 rpm, and depth of cut 0.25 mm, feed rate 0.1 mm/rev.

The highest consumption of active and reactive energy was obtained for parameters: spindle speed 1200 rpm, depth of cut 0.5 mm, feed rate 0.1 mm/rev.

As can be seen in Fig. 7, in most of the analyzed settings, active energy consumption is higher than reactive energy.

In two cases, the opposite is true, i.e. more passive than active energy is consumed. This case was observed during the machining with spindle rotational speeds of 1400 and 1600 rpm and depth of cut $ap = 0.5$ mm.

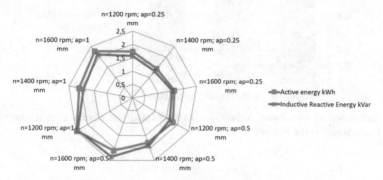

Fig. 7. Active and reactive energy consumption of selected machining parameters.

From the diagrams of active and reactive energy consumption presented in Figs. 8, 9 and 10 for the three spindle speeds, it can be stated:

– for depth of cut $ap = 0.25$ the lowest electricity consumption was observed for machining at a spindle speed of 1400 rpm (Fig. 8)
– for depth of cut $ap = 0.5$ the lowest electricity consumption was observed for machining at spindle speed of 1200 rpm (Fig. 9)
– for depth of cut $ap = 1$ the lowest electricity consumption was observed for machining at a spindle speed of 1400 rpm (Fig. 10)

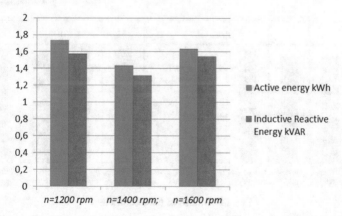

Fig. 8. Active and reactive power consumption for three spindle speeds and constant depth of cut $ap = 0.25$ mm feed $f = 0.1$ mm/rev.

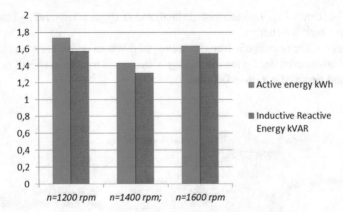

Fig. 9. Active and reactive power consumption for three spindle speeds and constant cutting depth $ap = 0.5$ mm feed $f = 0.1$ mm/rev.

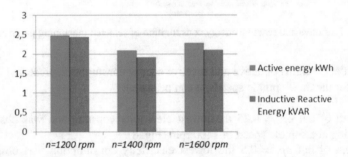

Fig. 10. Active and reactive power consumption for three spindle speeds and constant depth of cut ap = 1 mm and feed fn = 0.1 mm/rev.

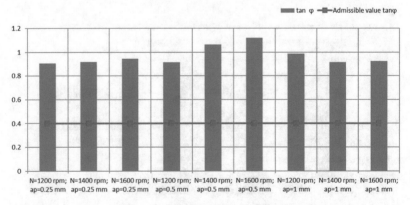

Fig. 11. Values of tanφ coefficient

The tanφ factor is very important for the companies, which do not have a reactive power compensation system, it pays for using it from the energy grid. This energy is consumed by the machine tools together with active electricity. For machine tools, inductive reactive energy costs will be incurred when the tanφ factor exceeds 0.4. There is no such threshold for capacitive reactive energy. However, as it results from conducted tests, machine tools draw mainly inductive reactive energy from the power grid. Hence, the value of tanφ indicator translates directly into costs incurred for electricity.

The largest values of *tan φ* coefficient were observed for machining with spindle speed at 1600 rpm and 1400 rpm and 0.5 mm depth of cut. The lowest values of the coefficient were obtained for machining at a spindle speed of 1200 rpm and a depth of cut of 0.5 mm and for machining at a cutting speed of 1400 rpm and a depth of cut of 1 mm. This relates directly to the costs of electricity charged by energy suppliers.

3 Summary

The analysis of the machining process, on the example of turning, shows how the use of various cutting parameters influences the use of electric energy, and thus the energy efficiency of the process.

The performed analysis points out that while considering the consumption of electric energy during the machining apart from use of active energy it is also necessary to analyze the use of reactive energy. This energy is also to a significant extent consumed by machine tool drives, such as active energy. The conducted research indicates that the choice of machining parameters may have an impact on the consumption of both active and reactive electricity.

Interesting is the fact that electricity consumption varies depending on the depth of cut and rotational speed. According to obtained data it is better machining with high speed and low depth of cut. The best energy efficiency was obtained for the spindle speed 1400 rpm and depth of cut 0.25 mm.

The obtained results confirm that, as in [9], by changing the processing parameters, we can have an impact on the reduction of energy consumption. In contrast to Diaz et al. [8] the results obtained confirm that the energy consumed not only depends on the machining time, but also on the machining parameters, especially the depth of cut and the rotational speed.

References

1. Li, W., Zein, A., Kara, S., Herrmann, C.: An investigation into fixed energy consumption of machine tools. In: Hesselbach, J., Herrmann, C. (eds.) Glocalized Solutions for Sustainability in Manufacturing, pp. 268–273. Springer, Heidelberg (2011)
2. Shin, S.-J., Woo, J., Rachuri, S.: Energy efficiency of milling machining: component modeling and online optimization of cutting parameters. J. Clean. Prod. **161**, 12–29 (2017)
3. Anand, Y., et al.: Optimization of machining parameters for green manufacturing. Cogent Eng. **3**, 1153292 (2016)
4. Chudy, R.: Wpływ parametrów technologicznych na energochłonność toczenia żeliwa sferoidalnego. Zesz. Nauk. Mech. Politech. Opol. **105**, 29–32 (2015)
5. Chudy, R., Grzesik, W.: Możliwości zmniejszenia energochłonności obróbki sekwencyjnej stali utwardzonej. Mechanik **89**(10), 1462–1463 (2016)

6. Grzesik, W., Żak, K.: Badanie wpływu zużycia ostrza na energię właściwą w dokładnym toczeniu stali 16MnCrS5 (AISI 5115). Mech. Miesięcznik Nauk.-Tech. (2015). http://www. mechanik.media.pl/artykuly/badanie-wplywu-zuzycia-ostrza-na-energie-wlasciwa-w-dok ladnym-toczeniu-stali-16mncrs5-aisi-5115.html. Accessed 18 Sept 2017
7. Niesłony, P., Żak, K., Chudy, R.: Ocena energetyczna wybranych parametrów stereometrii ostrza z CBN na kształtowanie powierzchni po toczeniu stali o podwyższonej twardości. Mechanik **89**(10), 1384–1386 (2016)
8. Diaz, N., Helu, M., Jarvis, A., Tönissen, S., Dornfeld, D., Schlosser, R.: Strategies for minimum energy operation for precision machining. In: The Proceedings of MTTRF 2009 Annual Meeting, UC Berkeley (2009)
9. Mori, M., Fujishima, M., Inamasu, Y., Oda, Y.: A study on energy efficiency improvement for machine tools. CIRP Ann. **60**, 145–148 (2011)
10. Liu, D., Wang, W., Wang, L.: Energy-efficient cutting parameters determination for NC machining with specified machining accuracy. Procedia CIRP **61**, 523–528 (2017). The 24th CIRP Conference on Life Cycle Engineering
11. Al Hazza, M.H.F., Adesta, E.Y.T., Riza, M., Suprianto, M.Y.: Power consumption optimization in CNC turning process using multi objective genetic algorithm. Adv. Mater. Res. **576**, 95–98 (2012)
12. Datta, R., Majumder, A.: Optimization of turning process parameters using multi-objective evolutionary algorithms. In: Proceedings of IEEE Congress on Evolutionary Computation, pp. 1–6 (2010)
13. Newman, S.T., Nassehi, A., Imani-Asrai, R., Dhokia, V.: Energy efficient process planning for CNC machining. CIRP J. Manuf. Sci. Technol. **5**, 127–136 (2012)
14. Wang, L., Wang, W., Liu, D.: Dynamic feature based adaptive process planning for energy-efficient NC machining. CIRP Ann. **66**, 441–444 (2017)
15. Zhang, H., Deng, Z., Fu, Y., Lv, L., Yan, C.: A process parameters optimization method of multi-pass dry milling for high efficiency, low energy and low carbon emissions. J. Clean. Prod. **148**, 174–184 (2017)
16. Ingarao, G.: Manufacturing strategies for efficiency in energy and resources use: the role of metal shaping processes. J. Clean. Prod. **142**, 2872–2886 (2017)
17. Terelak-Tymczyna, A., Pawlukowicz, P.: Wskaźniki wyniku energetycznego dla procesu obróbki skrawaniem na przykładzie toczenia. Mechanik **87**(8–9), CD2 (2014)
18. Terelak-Tymczyna, A., Miądlicki, K., Nowak, M.: Efektywność energetyczna procesu obróbki skrawaniem na przykładzie toczenia. Mechanik **89**(10), 1308–1309 (2016)

Quality of Gear Wheels in Selected Operations of the Manufacturing Process

Piotr Zyzak[1]([✉]), Paweł Kobiela[2], Marek Gabryś[2], and Agnieszka Gawlak[1]

[1] University of Bielsko-Biala, Bielsko-Biala, Poland
pzyzak@ath.bielsko.pl, gawlak.a@vp.pl
[2] BEFARED S.A. Company, Bielsko-Biala, Poland
{pawel.kobiela,marek.gabrys}@befared.pl

Abstract. The article presents results of a set of tests carried out to determine manufacturing accuracy of gear wheels in selected operations of the manufacturing process. Particular attention was paid to accuracy of the teeth after performed operation of hobbing, induction hardening and profile dividing grinding. The values of the deviations measured using the coordinate method and directly on the Rapid Höfler grinder, which characterize teeth of the gear wheels, as well as diagrams showing profile, tooth-line, pitch and runout are presented.

Keywords: Gear wheels · Technological quality · Values of deviations

1 Introduction

Issues of production of gear wheels, characterized by specified parameters, and upon strict technological requirements, have been currently analyzed by different research institutions, as well as by specialized enterprises dealing with production of the gear wheels. The BEFARED S.A. Company from Bielsko-Biala is one leading manufacturer of reduction gears and general purpose gear-motors. This company has to satisfy strict requirements, both in the design stage of gear transmissions, and especially of gear wheels with imposed high technological quality, as well as in the stages of manufacturing and assembly of the gear transmissions.

A correct technological process to achieve machining of the gear teeth belongs to an important success factor, to obtain failure- free and silent-running operation of the gear transmission. Hence, in case of the technological processes, it is necessary to search after relation between quality of a performed operation and the final effect of this operation in terms of compliance with assumed accuracy class of produced teeth. Factors directly resulting from accomplishment of the manufacturing process can't be neglected here. They are: condition of used cutting tools, implemented machine tool, knowledge of operators, correctness of performed processes, including turning, milling of teeth, heat treatment and thermochemical treatment, and as well as finishing operation and grinding of the teeth.

G. M. Królczyk et al. (Eds.): IMM 2019, LNME, pp. 85–96, 2020.
https://doi.org/10.1007/978-3-030-49910-5_9

The main objective of this paper is to describe the technological quality of teeth of the gear wheels after completion of some selected operations of the technological process: hobbing, induction hardening [25–27] and profile dividing grinding.

Gear hobbing is one of the most productive manufacturing processes for cylindrical gears. The quality of the gears is a result of the tool quality, the precision of the workpiece, tool clamping and kinematics of the machine [23, 24].

Gear hobbing is a generating process. The term generating refers to the fact that the gear tooth form cut is not the conjugate form of the cutting tool, the hob. During hobbing both the hob and the workpiece rotate in a continuous rotational relationship. During this rotation, the hob is typically fed axially with all the teeth being gradually formed as the tool traverses the work face [21, 22].

It can be stated that technological quality of the teeth after operations preceding profile dividing grinding will determine preparation of the grinding strategy, including number of passes and number of travels of grinding wheel, number of dressings of grinding wheel, and fastening method of the gear wheel in grinding machine. Machining time is determined by strategy of the machining, while in case of production of the workpieces in series, it will determine its cost, which belongs to important factors to analysis of production planning and control systems.

2 General Characteristics of Analysed Operations in Manufacturing Process of the Investigated Gear Wheels

We present the analysis of three selected operations from the technological process of the gear wheels, namely: hobbing, induction hardening and profile dividing grinding.

Milling is the most used profiling method of the gear wheels with external teeth. Among milling operations we can distinguish hobbing, formed cutter milling, gear cutter milling and shank cutter milling [14]. Main rotary motion is performed by a multi-edge cutting tool (mill or face milling cutter), while linear plane motion or curvilinear motion is usually performed by a table with the workpiece. The milling operation is used mainly for the machining of planes, shaped surfaces and grooves and threads [13, 14]. The advantages of the milling operation include very high productivity and accuracy, low roughness of machined surface, and possibility of machining of various surfaces. The BEFARED S.A. Company uses hobbing operation to produce the gear wheels. Course of the hobbing operation is presented in the Fig. 1.

Hardening was the next investigated operation in production process of the gear wheels. The hardening is the heat treatment consisting in heating the workpiece to a suitable temperature in order to produce austenitic structure of the material, soaking in this temperature, and next, rapid cooling–cooling down to temperature of martensitic structure. Results are higher hardness and strength, and also changed internal structure [16]. Parameters having effect on mechanical properties of the material are [16]:

- soaking temperature – depends on chemical constitution of the material, on carbon content mainly,
- heating time – depends on chemical constitution and size of the workpiece, type of a furnace, heating medium, and heating conditions,

Fig. 1. Scheme of the cutting operation of a teeth using hobbing method: 1- workpiece, 2- hobbing gear cutter [14, 15].

- cooling rate – depends on cooling medium. Water and hydrous solutions of chloride assure the highest cooling rate. Compressed air and oils assure lower rates. To avoid a cracks and deformations, selection of the cooling medium should be suitable to material structure obtained after the cooling, and to values of quenching stresses generated in case of too rapid cooling.

We can distinguish such methods of the hardening as [2, 16]:

- customary martensitic hardening – consisting in continuous cooling in medium, e.g. in water. It is the simplest method of hardening, but results in high quenching stresses,
- surface hardening – consists in rapid and intensive heating of material's surface, and next, also rapid and intensive cooling down, to avoid penetration of the heat inside the material. In such way only thin surface layer of the material is hardened, what increases hardness and abrasion resistance, while core of the material remains more soft and ductile. The surface hardening can be performed when carbon content is at least 0,35%, i.e. the material is suitable for the heat treatment. In case of the gear wheels, this method consists in hardening of flanks of the teeth (Fig. 2), or additionally, also hardening on bottom land of the tooth. It allows for increase of wear resistance in case of flanks of the tooth, and increase of bending strength in bottom area of the tooth in case of hardened bottom lands of the teeth.

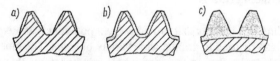

Fig. 2. Methods of surface hardening: a) hardening of flanks of the teeth, b) hardening of flanks and bottom lands of the teeth, c) circumferential hardening [2].

Simultaneous hardening of the both flanks is more advantageous than hardening of bottom lands of the teeth due to generated hardening strain. Scheme of induction hardening with method "tooth after tooth" of the teeth of gear wheel is shown in the Fig. 3.

Fig. 3. Operation of induction hardening of selected gear wheels in industrial conditions.

Profile-dividing grinding is the most accurate method of gear finishing machining, especially with unsymmetrical grinding, due to the more favorable working conditions of the tool [3, 9].

There are three methods of profile grinding [1, 3]:

- with symmetrical position of a grinding wheel – the advantage of this method is that two sides of adjacent teeth are ground at the same time; working conditions of the grinding wheel, resulting from the variable grinding depth along the profile of the grinding wheel during radial feed, can be regarded as a disadvantage;
- with asymmetrical position using one grinding wheel – the advantage of this method is more favorable grinding wheel working conditions, where only one side of the tooth is machined;
- with asymmetrical position of two grinding wheels – the advantage of this method is favorable grinding wheel operating conditions, and the disadvantage – the necessity of using two grinding wheels.

Profile grinding is a process that provides great potential with respect to flexibility and quality, and can by used to grind a wide variety of gears. The kinematics of profile grinding are much easier than in threaded wheel grinding since this process does not require so many synchronized movements. A profiled grinding wheel is used, witch is swiveled to the helix angle of the gear and rotates to perform the cutting speed. In addition, the wheel must be able to be moved in a radial direction to remove stock from the flanks, and in an axial direction along the gear axis [18–20].

It is assumed that the undoubted advantages of the profiling methods are: possibility of giving the outline of teeth of any shape and the possibility of grinding teeth in wheels with internal toothing [7–9].

Exemplary methods of profile dividing grinding of the internal and external teeth of the gear wheel are presented in the Fig. 4.

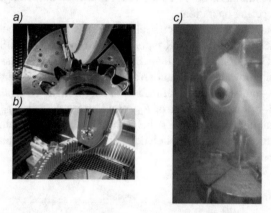

Fig. 4. Profile dividing grinding: a) external, b) internal, c) profiling of the grinding wheel.

3 Experimental Research

Within framework of the experimental research it has been decided to produce four gear wheels from the C55 steel, having the following chemical analysis: 0,52–0,6% C, 0,6–0,9% Mn, max 0,4% Si, max 0,045% P, max 0,045% S, max 0,4% Cr, max 0,4% Ni, max 0,1% Mo; parameters of the gear wheels were as follows:

- number of teeth $z = 33$,
- normal module $m = 5$,
- tip diameter of the teeth $da = 177,490$ mm,
- helix angle $\beta = 9°53'30''$,
- helix direction – LH.

A guidebook, describing sequence of implemented operations, was used to carry out the machining of the gear wheels. According to this guidebook [17], operation no. 60 corresponds to milling the teeth with hobbing method. Therefore, it has been adopted allowance for measurement over the teeth amounting to 0,3 mm. Teeth of the gear wheel no. 4 were cut with a gear-hobbing cutter on the SAMPUTENSILI S800 hobbing machine with the SINUMERIK nc unit. In course of the hobbing operation it have been used the following cutting parameters: $n_f = 186,09$ rpm $f = 2,50$ mm/rotation, $vc = 70$ m/min and intensive cooling and lubrication with base oil of MOBILMET 443 brand with improvers.

After hobbing and deburring of the gear teeth, they were subjected to heat treatment (induction hardening by tooth-to-tooth method to obtain a hardness of $54 \div 58$ HRC). Induction hardening of individual teeth was carried out using the feed method, where the operation is divided into one tooth during the operation. HF 35 kW generator and IV630 feeder were used at $300 \div 400$ kHz frequency and 670 A current, machine feed 2 and hardened layer thickness about 1 mm. After induction hardening, the wheels were subjected to a tempering operation.

After machining of teeth, the gear wheels were measured on 3D coordinate machine of the ZEISS PRISMO NAVIGATOR type, produced by CARL ZEISS, with scanning head of the Vast Gold brand, using the ZEISS GEAR PRO Involute 2014 measuring software. Maximal allowable deviation error of this machine amounts to MPE $= 2 + 3L$ μm; where: L – numerical value of measured length in m. The following deviations were measured [4–6]: profile total error F_α, profile form deviation $f_{f\alpha}$, profile angular deviation $f_{H\alpha}$, total helix deviation F_β, helix form deviation $f_{f\beta}$, helix slope deviation $f_{H\beta}$, pitch deviation f_p, cumulative pitch deviation, total F_p, single pitch difference f_u, radial runout of tooth space F_r.

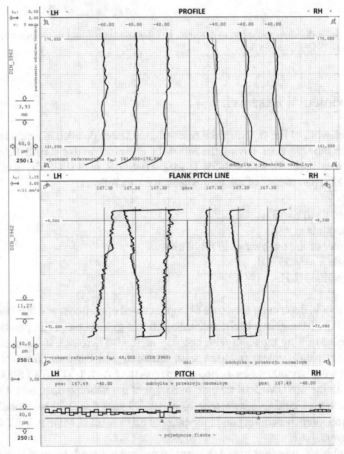

Fig. 5. Measurement results of profile and flank pitch line and pitch after hobbing operation- gear wheel no. 4.

The Fig. 5 shows diagrams depicting measurement results of the profile, of the flank pitch line of bearing surface of the tooth, and of the pitch of selected gear wheel no. 4, as have been obtained from the coordinate machine.

The same as in case of hobbing operation of the teeth, suitable measurements were performed for the gear wheels after induction hardening, using station of the coordinate machine of the ZEISS PRISMO NAVIGATOR type, produced by CARL ZEISS. Obtained measurement results for the gear wheel are presented in the Fig. 6.

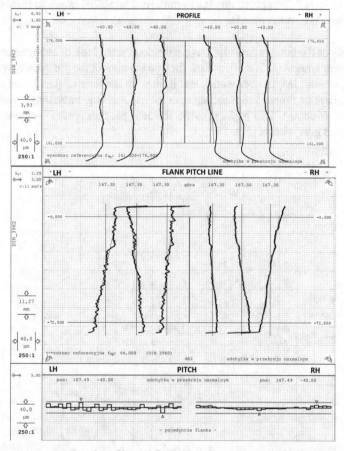

Fig. 6. Measurement results of profile and flank pitch line and pitch after induction hardening – gear wheel no. 4.

To perform profile dividing grinding operation, the following parameters had been given as input in the control program of the machining: radial approach of dressing, overlap index k_d [10–12], dressing parameter of grinding wheel η. For the investigated gear wheels, for the roughing and finishing operations the following parameters were adopted:

- pass 1: radial approach of dressing 0,05 mm, overlap index $k_d = 1,0$, $\eta = 0,6$, number of travels 6, approaching 0,025 mm, feed rate 7000 mm/min;

- pass 2: radial approach of dressing 0,05 mm, overlap index $k_d = 1,0$, $\eta = 0,6$, number of travels 6, approaching 0,025 mm, feed rate 7000 mm/min;
- pass 3: radial approach of dressing 0,05 mm, overlap index $k_d = 1,0$, $\eta = 0,6$, number of travels 6, approaching 0,025 mm, feed rate 7000 mm/min;
- pass 4: radial approach of dressing 0,05 mm, overlap index $k_d = 2,5$, $\eta = 0,4$, number of travels 4, approaching $2 \times 0,015$ mm and $2 \times 0,010$ mm, feed rate 6000 mm/min;
- pass 5: radial approach of dressing 0,05 mm, overlap index $k_d = 2,5$, $\eta = -0,5$, number of travels 2, approaching 0,010 mm, feed rate 2200 mm/min (first travel) and 1800 mm/min (second travel).

Analysis of the obtained results was performed on the base of changes occurring in particular gear wheels (1, 2, 3, 4), at first, then on the base of the parameter type (profile, line, pitch, runout, length over teeth), and in turn on the base of performed machining operations, heat treatment and finishing operations (hobbing, induction hardening, profile dividing grinding). This paper presents the analysis concerning only one out of the four produced gear wheels (Fig. 7).

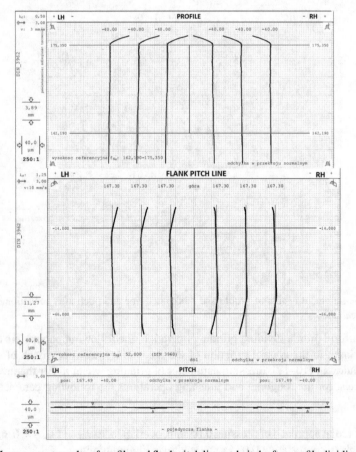

Fig. 7. Measurement results of profile and flank pitch line and pitch after profile dividing grinding – gear wheel no.4.

4 Discussion of Obtained Results

To assess quality of the profile of produced gear wheel, the following deviations have been used: profile total error F_α, profile form deviation $f_{f\alpha}$ and profile angular deviation $f_{H\alpha}$.

After hobbing operation the results are as follows:

- LH profile of bearing surface of the tooth: the 7th class of accuracy has been confirmed for all deviations,
- RH profile of bearing surface of the tooth: limits for the 7th class were exceeded for all deviations. Deviation of profile total error F_α and profile angular deviation $f_{H\alpha}$ have obtained class 9, while profile form deviation $f_{f\alpha}$ has obtained class 8.

After induction hardening operation the results are as follows:

- LH profile of bearing surface of the tooth: the 7th class of accuracy has been fulfilled for the deviation of profile total error F_α and profile angular deviation $f_{H\alpha}$, while profile form deviation $f_{f\alpha}$ has obtained a lower class, i.e. 8th class,
- RH profile of bearing surface of the tooth: the 7th class of accuracy has been exceeded in case of all deviations. Deviation of profile total error F_α has obtained 9th class, while profile form deviation $f_{f\alpha}$ and profile angular deviation $f_{H\alpha}$ have obtained class 8.

After performed operation of profile dividing grinding the gear wheel was measured by control-measuring device of the grinder and by the coordinate machine. Results of measured deviations are summarized in the Table 1.

Table 1. Values of obtained deviations of profile of bearing surface of the tooth $F_\alpha, f_{f\alpha}, f_{H\alpha}$ after profile dividing grinding operation.

LH profile	Rapid Höfler grinder			ZEISS PRISMO NAVIGATOR		
	F_α	$F_{f\alpha}$	$F_{H\alpha}$	F_α	$F_{f\alpha}$	$F_{H\alpha}$
Teeth 1	2,5/9 μm	2,5/7 μm	−1,5/±5,5 μm	5/18 μm	2/14 μm	−5/±10 μm
Teeth 12	3,5/9 μm	3/7 μm	0,5/±5,5 μm	4/18 μm	2/14 μm	−5/±10 μm
Teeth 23	3,5/9 μm	3,5/7 μm	−0,5/±5,5 μm	3/18 μm	1/14 μm	−3/±10 μm
$\bar{x}$	3,2/9 μm	3/7 μm	−0,5/±5,5 μm	4/18 μm	1/14 μm	−4/±10 μm
RH profile	Rapid Höfler grinder			ZEISS PRISMO NAVIGATOR		
	F_α	$f_{f\alpha}$	$f_{H\alpha}$	F_α	$f_{f\alpha}$	$f_{H\alpha}$
Teeth 1	4/9 μm	3/7 μm	1,5/±5,5 μm	3/18 μm	2/14 μm	−3/±10 μm
Teeth 12	4/9 μm	3/7 μm	2/±5,5 μm	2/18 μm	2/14 μm	−2/±10 μm
Teeth 23	3,5/9 μm	3,5/7 μm	0,5/±5,5 μm	2/18 μm	2/14 μm	−1/±10 μm
$\bar{x}$	3,8/9 μm	3,2/7 μm	1,3/±5,5 μm	3/18 μm	2/14 μm	−2/±10 μm

To assess machining quality of flank pitch line of bearing surface of the tooth, the following deviations are used: total helix deviation F_β, helix form deviation $f_{f\beta}$ and helix slope deviation $f_{H\beta}$.

After hobbing operation the results are as follows:

- LH flank pitch line of bearing surface of the tooth: the 7th class of accuracy was exceeded by all deviations. Helix form deviation $f_{f\beta}$ has obtained 8th class of accuracy, while in case of other two deviations, i.e. total helix deviation F_β and helix slope deviation $f_{H\beta}$ they obtained class 10,
- RH flank pitch line of bearing surface of the tooth: the 7th class of accuracy has been fulfilled for the parameter $f_{f\beta}$ only, and achieved a higher, 5th class, while in case of two other deviations of F_β and $f_{H\beta}$ this class has been exceeded and has reached class 10.

After induction hardening the results are as follows:

- LH flank pitch line of bearing surface of the tooth: the 7th class of accuracy was exceeded by all deviations. total helix deviation F_β has obtained 10th class, helix form deviation $f_{f\beta}$ has obtained class 8, while helix slope deviation $f_{H\beta}$ has obtained class 11,
- RH flank pitch line of bearing surface of the tooth: the 7th accuracy class was fulfilled in case of helix form deviation $f_{f\beta}$ only, reaching class 6. Total helix deviation F_β and helix slope deviation $f_{H\beta}$ have obtained class 10.

The following results have been obtained after of profile dividing grinding operation (Table 2):

Table 2. Values of obtained deviations of flank pitch line of bearing surface of the tooth $F_\beta, f_{f\beta}, f_{H\beta}$ after of profile dividing grinding operation.

LH flank pitch line	Rapid Höfler grinder			ZEISS PRISMO NAVIGATOR		
	F_β	$F_{f\beta}$	$F_{H\beta}$	F_β	$F_{f\beta}$	$F_{H\beta}$
teeth 1	10/10 μm	9,5/7 μm	−1,5/±7 μm	5/18 μm	4/12 μm	6/±14 μm
teeth 12	4,5/10 μm	4,5/7 μm	0/±7 μm	3/18 μm	4/12 μm	2/±14 μm
teeth 23	5/10 μm	4/7 μm	−1,5/±7 μm	4/18 μm	4/12 μm	4/±14 μm
$\bar{x}$	6,5/10 μm	6/7 μm	−1/±7 μm	4/18 μm	4/12 μm	4/±14 μm
RH flank pitch line	Rapid Höfler grinder			ZEISS PRISMO NAVIGATOR		
	F_β	$f_{f\beta}$	$f_{H\beta}$	F_β	$f_{f\beta}$	$f_{H\beta}$
Teeth 1	5,5/10 μm	5,5/7 μm	0/±7 μm	5/18 μm	6/12 μm	3/±14 μm
Teeth 12	4,5/10 μm	4,5/7 μm	−0,5/±7 μm	6/18 μm	6/12 μm	−1/±14 μm
Teeth 23	5,5/10 μm	5,5/7 μm	−0,5/±7 μm	6/18 μm	6/12 μm	3/±14 μm
$\bar{x}$	5,2/10 μm	5,2/7 μm	−0,3/±7 μm	6/18 μm	6/12 μm	2/±14 μm

To assess machining quality of pitch value of the gear wheel, the following deviations are used: cumulative pitch deviation, total F_p single pitch difference f_u.

After hobbing operation and induction hardening, accuracy classes of pitch deviations were included between classes $5 \div 8$ respectively.

Runout of the gear wheels is assessed by deviation of radial runout of tooth space F_r. Results are summarized in the Table 3.

Table 3. Values of accuracy classes and deviations of radial runout F_r.

	Accuracy classes	F_r
After hobbing	8/7	42/36 μm
After induction hardening	8/7	42/36 μm
After profile dividing grinding-Höfler machine	2/5	6/18 μm
After profile dividing grinding-coordinate machine	5/7	12/36 μm

5 Summary

On the base of the performed investigations, aimed at the assessment of technological quality of teeth of the gear wheels, after performance of selected operations in the process of their manufacturing, it has been ascertained that:

- after hobbing operation the deviations which characterize profile of the tooth ($F_\alpha, f_{f\alpha}, f_{H\alpha}$) meet the requirements of the 8th and 9th accuracy class of manufacturing, while deviations of the flank pitch line of bearing surface of the tooth ($F_\beta, f_{f\beta}, f_{H\beta}$) meet the requirements of the 10th and 11th accuracy class of manufacturing;
- after induction hardening with "tooth after tooth" method, where nearly all deviations have been substantially exceeded with respect to values allowable by the standard, and hence, deviations of the profile are within 9th class, while deviations of the flank pitch line of bearing surface of the tooth are within 11th accuracy class of manufacturing;
- profile dividing grinding of the investigated gear wheels was considered as finishing operation. On the base of obtained results of measurements accomplished on the machine tool directly after completion of the machining operations of each from the gear wheels, and next performed on the coordinate machine, it has been confirmed that manufacturing quality of teeth of the gear wheels corresponds to the 5th class of accuracy, whereas objective as the 7th accuracy class was assumed in the machining program.

Design of manufacturing processes, especially in case of the gear wheels, requires great knowledge and acquaintance of processes and phenomena determining required condition of the workpiece at a given stage of the manufacturing.

Results obtained after finishing operation, as well as the way this operation is performed, will directly result from the accuracy obtained from the performed preceding operations such as hobbing and induction hardening.

References

1. Oczoś, K.E, Porzycki, J.: Grinding. Basics and technology. WNT, Warsaw (1986)
2. Ochęduszko, K.: Gears. T. II. Execution and assembly. WNT, Warsaw (1992)
3. Oczoś, K.E., Marciniak, M.: Development of machine tool construction for grinding processes Part 2. Mechanik **79**(3), 192–198 (2006)
4. PN-ISO 1328-1: 1997: Helical gears. Accuracy of implementation by ISO. Definitions and values of deviations of the unilateral tooth sides
5. PN-ISO 1328-2: 1997: Definitions and values of composite measurement deviations and runout deviations
6. DIN 3962: 1978: Toleranze für Stirnradverzahnungen. Toleranze für abweichungen einzelner Bestimmungsgrößen
7. User manual for the Rapid 2000 shape-sensitive grinding machine for finishing the teeth of Höfler, Germany
8. www.klingelnberg.com/en/business-divisions/hoefler/cylindrical-gear-grinding-machines/detail-page/product/rapid-2000/. Accessed 20 Apr 2019
9. Płonka, S., Szadkowski, J., Matuszek, J., Kobiela, P.: Development of selected methods of machining gear teeth. Zeszyty Naukowe Politechniki Śląskiej. Seria Transport **87**(1929), 11–20 (2015)
10. Rosik, R., Świerczyński, J.: Influence of the MQL method on the parameters of shaping the active surface of the grinding wheel and the roughness of the surface layer of the workpiece. Inżynieria Maszyn **16**(1–2), 175–185 (2011)
11. Zyzak, P., Kobiela, P., Brożek, A., Gabryś, M.: The influence of the grinding strategy on the accuracy and roughness of gear teeth. Mechanik **91**(8–9), 737–740 (2018)
12. Płonka S., Zyzak P., Kobiela P., The effect of dressing the grinding wheel on the precision of the teeth grinded using a shaping method, Zeszyty Naukowe Politechniki Rzeszowskiej Mechanika, t. XXXIV **89**(4/17), 537–546 (2017)
13. Storch, B.: The basics of machining. Koszalin (2001)
14. Zaleski, K., Matuszak, J.: Basics of waste treatment. Politechnika Lubelska, Lublin (2016)v
15. http://virt.com.pl/p/8/68/frezowanie-obwiedniowe-kol-walcowych-uslugi-kooperacyjne. html. Accessed 10 May 2018
16. https://mech.pg.edu.pl/documents/174709/…/PIM_5_hartowanie_odpuszczanie.pdf. Accessed 14 Feb 2018
17. Technology Operations Guidebook D-U09-280-Z1-33, FRiM BEFARED S.A. (2000)
18. Richmond, D.: CNC gear grinding methods. Gear Technol. J. Gear Manuf., 43–50 (1997)
19. Türich, A.: Producing profile and lead modifications in threaded wheel and profile grinding. Gear Technol. J. Gear Manuf., 54–62 (2010)
20. Türich, A., Kobialka, C., Vucetic, D.: Innovative concepts for grinding wind power energy gears. Gear Technol. J. Gear Manuf., 39–44 (2009)
21. Komori, M., Sumi, M., Kubo, A.: Simulation of hobbing for analysis of cutting edge failure due to chip crush. Gear Technol. J. Gear Manuf., 64–69 (2004)
22. Gimpert, D.: The gear hobbing process. Gear Technol. J. Gear Manuf., 38–44 (1998)
23. Gravel, G.: Simulation of deviations in hobbing and generation grinding. Gear Technol. J. Gear Manuf., 56–60 (2014)
24. Krömer, M., Sari, D., Löpenhaus, C., Brecher, C.: Surface characteristics of hobbed gears. Gear Technol. J. Gear Manuf., 68–75 (2017)
25. Rudnev, V.: Trends in induction hardening. Gear Technol. J. Gear Manuf., 22–23 (2019)
26. Li, Z.: Innovative induction hardening process with pre-heating for improved fatigue performance of gear component. Gear Technol. J. Gear Manuf., 62–68 (2014)
27. Rudnev, V.: Recent inventions and innovations in induction hardening of gears and gear-like components. Gear Technol. J. Gear Manuf., 66–69 (2013)

Analysis of Optimization Effects of Deflection Angle of a Disc-Type Grinding Wheel on Accuracy of Profile Dividing Grinding of Gears

Piotr Zyzak[1]([✉]), Paweł Kobiela[2], Marek Gabrys[2], and Agnieszka Gawlak[1]

[1] University of Bielsko-Biala, Bielsko-Biala, Poland
pzyzak@ath.bielsko.pl, gawlak.a@vp.pl
[2] BEFARED S.A. Company, Bielsko-Biala, Poland
{pawel.kobiela,marek.gabrys}@befared.pl

Abstract. The article presents results of the research on grinding operations of a gear wheels, grinded with use of profile dividing method on the Rapid Höfler grinders, with consideration of deflection angle of the grinding wheel. Essential investigations had concerned effect of the helix angle β of the grinding wheel on accuracy of grinded teeth. Results of the research are presented in tabular form, where reference was done to obtained values of the deflections characterizing grinded gears, and in graphical form with diagrams. The most accurate results of the measurements were obtained on the gear wheels marked as 1 and 4; 74% of the accuracy classes of these gears have obtained classes 1–3, and other 27% obtained classes between 4–5.

Keywords: Gear wheel · Profile dividing grinding · Optimization

1 Introduction

Analysis of publications and specialized elaborations dealing with designing, manufacturing, and especially grinding of the gear wheels is pointing at the grinding operation as still the leading technique to produce gear wheel with predetermined characteristic of the teeth. We can distinguish two grinding methods of the teeth: profile grinding method [1, 2, 21] and generation grinding method [13, 14].

Nowadays, many gears need to be hard finished. Continuous generating grinding offers a very high process efficiency. For these reasons, the gear production is always looking for hard finishing method, witch is, on one hand, capable of creating all possible gear modifications, and on the other hand, as efficient as possible [13, 15]. In the generating method of gear grinding, the shape of gear tooth is the results of combined movements of the work piece and the grinding wheel, witch is usually dressed to the shape of a rack tooth [14, 18].

Profile grinding is a process that provides great potential with respect to flexibility and quality, and can by used to grind a wide variety of gears. The kinematics of profile

© Springer Nature Switzerland AG 2020
G. M. Królczyk et al. (Eds.): IMM 2019, LNME, pp. 97–108, 2020.
https://doi.org/10.1007/978-3-030-49910-5_10

grinding are much easier than in threaded wheel grinding since this process does not require so many synchronized movements. A profiled grinding wheel is used, witch is swiveled to the helix angle of the gear and rotates to perform the cutting speed. In addition, the wheel must be able to be moved in a radial direction to remove stock from the flanks, and in an axial direction along the gear axis. If the gear is helical, it requires a continuous rotational movement in order to follow the lead and a discontinuous pitch movement to grind all teeth [22, 23].

In tooth profile grinding, the stock removal depends on the chosen process strategy [19, 20].

Analysis of the profile dividing process in grinding operation of a gear wheel's teeth performed within industrial conditions enables to state that preceding operations (milling, heat treatment, and thermochemical treatment), accuracy of positioning, method of attachment of the teethed element to grinding operation, and implemented grinding wheel, belong additionally to conditions which determine final result of the machining. Developed each time strategy how to perform the profile dividing grinding takes into considerations a shaping method of working profile of the grinding wheel (number of dressing operations), and parameters of the machining with consideration of number of passes and travels of the grinding wheel in process of the machining [16, 17]. It is assumed that to advantages of the profiling methods belong possibility of arbitrary shaping of profile of tooth, and possibility of grinding gear wheels with internal teeth [7, 9].

Use of such machining enables obtainment of predefined properties, understood as characteristics, and expressed directly as suitable deviations, which should be measured directly on 3D measuring machine. Additionally, assessment of surface roughness parameters of the teeth needs to be performed after the grinding operation. Among the most important deviations, the following ones can be distinguished [4–6]: profile total error F_α, profile form deviation $f_{f\alpha}$, profile angular deviation $f_{H\alpha}$, total helix deviation F_β, helix form deviation $f_{f\beta}$, helix slope deviation $f_{H\beta}$, pitch deviation f_p, cumulative pitch deviation, total F_p, single pitch difference f_u, radial runout of tooth space F_r.

2 Characteristic of Workstation with Rapid Höfler 2000 Grinder

The Rapid Höfler 2000 grinder to profile grinding of a gear wheels is installed in the Befared S.A. Company, this Company belongs to a leading manufacturers of reduction gears and gear-motors of a general purpose. In the Fig. 1 is shown a view of the machine tool, comprising description of its axes together with main movements along the axes, and profiling course of the grinding wheel. The axes are assigned to the main subassemblies, which perform the following functions [7, 8]: wheelhead (axis A) – deflection of the grinding wheel to the angle of inclination, workpiece's table (axis B) - pitch, grinding spindle (axis C) – rotary motion of grinding spindle (axis C2) – rotary motion of dressing roller during dressing operations, slides for workpiece (axis X) – adjustment of diameter, slides of tool (axis Y) – stepwise motion, dressing slides (axis Z2) – generation of profile of the tool. Important technical data of the workpiece are as follows: tip diameter of the gear $da = 2000$ mm (max.), module pitch (depending on parameters of the teeth) $m = 3,5 \div 30$ mm and maximal inclination angle LH/RH 45°.

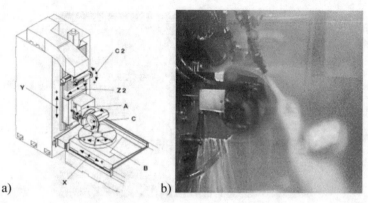

a)   b)

Fig. 1. Rapid Höfler grinder: a) description of the axes: wheelhead (axis A) – deflection of the grinding wheel to the angle of inclination, workpiece's table (axis B) – pitch, grinding spindle (axis C) – rotary motion of grinding spindle (axis C2) – rotary motion of dressing roller during dressing operations, slides for workpiece (axis X) – adjustment of diameter, slides of tool (axis Y) – stepwise motion, dressing slides (axis Z2) – generation of profile of the tool, b) profiling of the grinding wheel [7, 9]

3   Profile Dividing Grinding of A Selected Gear Wheels with Consideration of Deflection Angle Optimization of the Grinding Wheel

Elaboration of data to the manufacturing task belong to essential elements connected with accomplishment of grinding operation of the teeth with use of profile dividing method on numerically controlled Rapid Höfler grinding machine. Detailed parameters of the machining task are input by an operator to special Gear Pro software, loaded into the control unit of the machine tool. Selecting a suitable, so called soft pushbuttons, there is a possibility to switch over areas of in grinding program. Data from a given order comprise all information indispensable to machining of a workpiece. Each order comprises its own data concerning the workpiece and the processes. After selection from list of required orders, there is a possibility to printout data of the order. It should be underlined that data to be input depend on options selected during creation of the order. This is directly connected with type of a grinded teeth: i.e. internal or external one. Among data of the task are distinguished processes which are understood as operations related directly to the order, and which are memorized together with them. The data and conditions of the processes to be performed later in manufacturing area are loaded to the input window. Optimization of the deflection angle of the grinding wheel is taken into considerations by the data of the order. In the Fig. 2 is presented view of the dialog window, where in its upper field are defined parameters of the machining, as required by the order; while in the bottom field are defined optimization parameters of the angle.

During grinding operation of the teeth of gear wheels the grinding wheel is deflected, as a standard, with bevel angle. In some situations set-up of a different rotation angle can be justified, e.g. in case of:

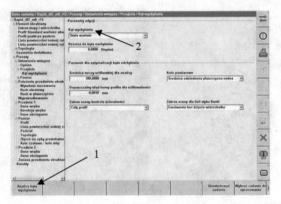

Fig. 2. View of the dialog window - Rapid Höfler 2000 grinding machine: 1- analysis of deflection angle, 2- the deflection angle β of the grinding wheel.

- generation of asymmetric modifications of side profile,
- minimization of over travel (it amounts to shortening of the grinding time),
- equalization of forces during entering and coasting of the grinding wheel (it amounts to presence of uniform pressure on teeth from RH and LH flank),
- height of contact line on flank (in case when one flank of the tooth is grinded).

The following options of the deflection angle of the grinding wheel can be distinguished:

- constant value – rotation angle can be input as a constant value. In case when such option was selected, a window to input difference with respect to the bevel angle is displayed,
- automatically optimized – rotation angle of the grinding wheel will be optimized automatically,
- optimized with respect to shift on measuring circumference – such optimization causes that the grinding operation is running simultaneously at similar height on the both flanks,
- optimized with respect to length of the travel – such optimization results in simultaneous contact of the grinding wheel with the workpiece on the both flanks. Owing to this, it is possible to minimize radial forces and optimize measurement diagram of side profile,
- optimized with respect to height of contact line – such optimization reduces deviations in grinding of a single flank.

When deciding for the analysis of rotation angle of the grinding wheel, the program pushbutton "Analiza kąta wychylenia" (Fig. 2) (Analysis of Deflection Angle) is used. Selection of this option allows the operator to analise rotation angle of the grinding wheel and its optimization on the base of the following criteria: minimal coasting of the travel, minimal shift at measuring circle, minimal height of contact line on bearing surface of tooth, preset value of the deflection angle. View of the window to optimization of the deflection angle of the grinding wheel is shown in the Fig. 3.

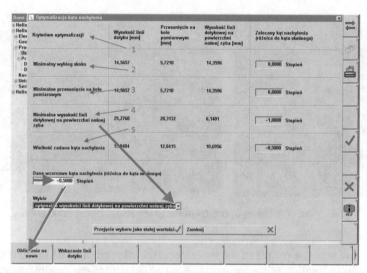

Fig. 3. View of the window "Analysis of the deflection angle": 1- optimization criteria, 2- minimal coasting of the travel, 3- minimal shift at measuring circle, 4- minimal height of contact line on bearing surface of tooth, 5- preset value of the deflection angle.

To grinding operation on grinding machine of the RAPID 2000 type produced by Höfler Company were subjected the gear wheels made of C55 steel with the following chemical analysis: 0,52–0,6% C, 0,6–0,9% Mn, max 0,4% Si, max 0,045% P, max 0,045% S, max 0,4% Cr, max 0,4% Ni, max 0,1% Mo, after induction hardening; and with the following parameters: $m = 5$ mm, $z = 33$, $dp = 167,490$ mm, $\alpha = 20°$, $\beta = 9°53'30''$, $x = 0$. Measurements of the gear wheels after grinding operation with different criteria of the optimization were performed directly on the workstation of the grinding machine. In the next step, measurements were performed on the 3D coordinate machine of ZEISS PRISMO NAVIGATOR type, produced by CARL ZEISS Company, and quipped with the Vast Gold scanning type measuring head, and using the ZEISS GEAR PRO Involute 2014 software for the deviations of: F_α, $f_{f\alpha}$, $f_{H\alpha}$, F_β, $f_{f\beta}$, $f_{H\beta}$, F_p, F_r [4–6]. Maximal allowable deviation error of this machine amounts to MPE = 2 + 3L μm; where: L – numerical value of measured length in m [11, 12].

After operations of: turning with allowance, toughening and turning, teeth of the gear wheel marked as no. 4 were-cut by hobbing cutter on a hobbing machine of the SAMPUTENSILI S800 type with SINUMERIK nc system. For the hobbing operation the following cutting parameters were used: $n_f = 186,09$ rpm, $f = 2,50$ mm/rotation and $vc = 70$ m/min, together with intensive cooling and lubrication with base oil of MOBILMET 443 type with improvers.

After milling and deburring operations of the gear wheel, the teeth were subjected to induction hardening, tooth after tooth, to hardness of 58–60 HRC. In the Fig. 4 are shown a view of one from the gears selected to investigations, fastened in the grinder, and course of the grinding operation.

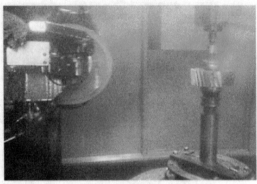

Fig. 4. View of grinding operation of the gear - Rapid Höfler 2000 grinder.

To assure accurate results of the measurements, after grinding of the first from the gear wheels, calibration of the measuring head of the control-measurement unit of the grinder was performed (Fig. 5).

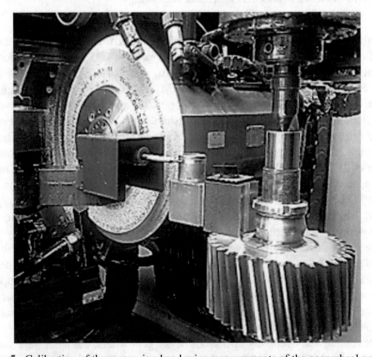

Fig. 5. Calibration of the measuring head prior measurements of the gear wheel no. 2.

Measurements of characteristic geometrical deviations of the teeth were performed with use of the measuring head (Fig. 6). In addition, evaluated corrections were input to the program.

Fig. 6. Measurements of the gear wheel no. 2 - Rapid Höfler 2000 grinder.

4 Results of Research and Discussion of Results

Essential investigations had concerned effect of the helix angle β of the grinding wheel on accuracy of grinded teeth. The accuracy can be presented in form of values of deviations, or in form of accuracy classes. To have the considerations legible, decision was taken to describe accuracy of the grinded teeth with use of the accuracy classes of manufactured gears. In case of the investigated gear wheels, it have been developed suitable machining programs, in which a value expressing difference in position of the grinding wheel with respect to assumed deflection angle is obtained directly after adoption of the criterion to optimization of the deflection angle of the grinding wheel. It was assumed to denominate such difference as δ. To obtain representative results for the gear wheels marked as 1, 2 and 4 respectively, a suitable-machining programs were developed, which for each from the gears directly output a different values of the parameter, described as difference with respect to deflection angle of the grinding wheel. However, the gear wheels No. 1 and 3 were produced using the same value of the parameter, described as difference to the deflection angle. Values of the parameter δ for the deflection angle in the gear wheels 1, 2 and 3 were negative, while for the gear wheel 4 were positive. Value of the deflection angle and the parameter δ for all gear wheels is presented in the Table 1.

Table 1. Values of the parameter δ for assumed criteria of optimization.

	Gear wheel 1	Gear wheel 2	Gear wheel 3	Gear wheel 4
Helix angle β	−9,8917°	−9,8917°	−9,8917°	−9,8917°
Deflection angle	Constant value	Constant value	Constant value	Constant value
Difference δ	−0,9142°	−0,1694°	−0,9142°	0,1266°

Completion of the grinding operations of the selected gear wheels has required input to the machining program of: radial approach angle of dressing, teeth overlap index k_d [10], parameter of dressing of the grinding wheel η (η – ratio of tangential velocity of dressing roller to tangential velocity of the grinding wheel). For the investigated gear wheels, the following parameters were taken to the roughing and finishing operations:

- pass 1: radial approach of dressing 0,05 mm, overlap index $k_d = 1,0$, $\eta = 0,6$, number of travels 6, approaching 0,025 mm, feed rate 7000 mm/min;
- pass 2: radial approach of dressing 0,05 mm, overlap index $k_d = 1,0$, $\eta = 0,6$, number of travels 6, approaching 0,025 mm, feed rate 7000 mm/min;
- pass 3: radial approach of dressing 0,05 mm, overlap index $k_d = 1,0$, $\eta = 0,6$, number of travels 6, approaching 0,025 mm, feed rate 7000 mm/min;
- pass 4: radial approach of dressing 0,05 mm, overlap index $k_d = 2,5$, $\eta = 0,4$, number of travels 4, approaching $2 \times 0,015$ mm and $2 \times 0,010$ mm, feed rate 6000 mm/min;
- pass 5: radial approach of dressing 0,05 mm, overlap index $k_d = 2,5$, $\eta = -0,5$, number of travels 2, approaching 0,010 mm, feed rate 2200 mm/min (first travel) and 1800 mm/min (second travel).

Due to large number of the data, selected results obtained for the gear wheel no. 2, grinded with use of profile dividing method with criterion of the optimization as minimal coasting of the travel, were present in form of appropriate tables and diagrams.

After performed profile dividing grinding operation, the gear wheel no. 2 was measured by the control-measurement unit of the grinding machine, and by the 3D coordinate machine. Obtainment of 5th class was set in the Rapid Höfler grinder, while 7th class of manufacturing was set in the 3D coordinate machine. It has been also considered as appropriate to present graphical runs of the profile, profile of bearing surface of the tooth and the pitch values for selected gear wheel marked as no. 2. Results obtained from the measurements on 3D coordinate machine are shown in the Fig. 7.

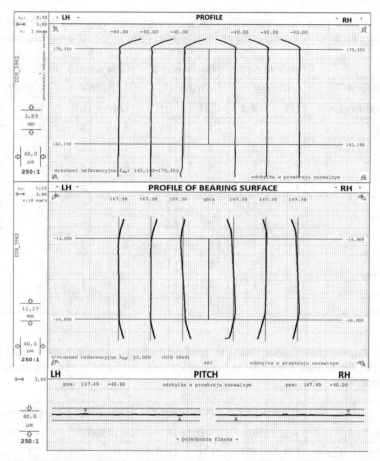

Fig. 7. Chart with measurement results of 3D measured by the 3D coordinate machine of ZEISS PRISMO NAVIGATOR type, produced by CARL ZEISS Company, and quipped with the Vast Gold scanning type measuring head, and using the ZEISS GEAR PRO Involute 2014 software.

Issues of grinding the teeth of a gear wheels are very complicated. In case of performed investigations, it has been accomplished assessment of profile dividing grinding accuracy for four optimization criteria of the deflection angle of the grinding wheel, and detailed results were elaborated.

In the Tables 2 and 3 are set up results illustrating obtained accuracy classes of the deviations for four gear wheels taken to the investigations, in reference to RH and LH flank of the tooth.

The best results in terms of measured deviations and corresponding accuracy classes of the machining were obtained for the gear wheel marked as no. 2, where parameter δ was the lowest. Measured deviations were included mainly in the class 1. Deviation of the radial run was classified in the lowest class, and was included in class 3. A little bit worse results were obtained for the gear wheel marked as no. 4, where deviation from the deflection angle was positive, as the only parameter. The deviations concerning profile of

Table 2. Accuracy classes corresponding to measured deviations, adequately to RH and LH flank of the tooth, for assumed criteria to the optimization measured by the 3D coordinate machine

Deviations/accuracy	Gear wheel 1		Gear wheel 2		Gear wheel 3		Gear wheel 4	
Helix angle β	$-10,8059°$		$-10,0611°$		$-10,8059°$		$-9,7651°$	
	LH	RH	LH	RH	LH	RH	LH	RH
F_α	3	2	3	1	4	3	3	2
$f_{f\alpha}$	2	2	1	1	1	1	1	2
$f_{H\alpha}$	5	1	4	1	6	5	5	3
F_β	4	3	6	4	4	5	3	4
$f_{f\beta}$	1	1	1	2	2	6	3	5
$f_{H\beta}$	5	5	7	6	6	3	5	2
f_p	1	1	2	2	2	2	2	2
F_p	3	2	1	1	4	4	4	3
f_u	1	1	2	1	1	1	1	2
F_r	5		4		5		5	

Table 3. Acuracy classes corresponding to measured deviations, adequately to RH and LH flank of the tooth, for assumed criteria to the optimization measured by the control-measurement unit of the grinding machine

Deviations/accuracy	Gear wheel 1		Gear wheel 2		Gear wheel 3		Gear wheel 4	
Helix angle β	$-10,8059°$		$-10,0611°$		$-10,8059°$		$-9,7651°$	
	LH	RH	LH	RH	LH	RH	LH	RH
F_α	2	2	1	2	1	1	2	3
$f_{f\alpha}$	2	2	2	2	1	2	3	3
$f_{H\alpha}$	1	1	1	1	1	1	1	2
F_β	6	6	1	1	6	6	5	3
$f_{f\beta}$	6	6	1	2	7	7	7	4
$f_{H\beta}$	4	4	1	1	4	5	1	1
f_p	1	1	1	1	1	1	1	1
F_p	1	1	1	1	2	1	1	1
f_u	1	1	1	1	1	2	1	1
F_r	1		3		2		2	

the teeth were included between classes 1–3, while deviations of the pitch have obtained class 1, deviation of the radial run has obtained class 2. Lower accuracy classes were obtained in case of helix deviations, and were ranged between classes 3–5. A special attention should be paid to exceeded accuracy class in case of LH side of deviation of helix form, which obtained class 7. In spite of higher accuracy classes obtained by total helix deviation and deviation of helix form (3–7), the profile angular deviation has obtained class 1.

Making analysis of results of the deviation classes measured on the 3D coordinate machine, it can be noticed that these deviations are different that the ones measured on the control-measurement unit of the grinding machine. Deviations of none from the gear wheels, in majority of the cases, were not included within accuracy classes 1–2. The most accurate results of the measurements were obtained on the gear wheels marked as 1 and 4; 74% of the accuracy classes of these gears have obtained classes 1–3, and other 27% obtained classes between 4–5. Simultaneously, sizes of the accuracy classes are not related with each other in any way, and do not show any schematics of occurrence. In addition, obtained results of accuracy classes of the gear wheels marked as 1 and 3, which were grinded with the same difference with respect to the deflection angle, differ from each other diametrically in part of the results. The main difference is number and size of accuracy classes of the deviations. In case of the gear wheel no. 3, only 58% of the deviations were between 1–3 class of the accuracy, 32% deviations were between 4–5 class, and 16% of the deviations have obtained class 6. Similar to results of the gear wheel no. 3 are results of the gear wheel no. 2. Only 63% of results of the deviations have received classes between 1–3; 16% of results of the deviations received classes between 6–7, and remaining results were in class 4. Low accuracy class can be observed in case of total helix deviation F_β and helix slope deviation $f_{H\beta}$.

In course of assessment of the results from performed experimental investigations attention should be paid to the fact that differences between high accuracy classes (1–3) amount from one to a few micrometers.

5 Summary

Preparing a task of grinding operation of gear wheels using the profile dividing method requires high knowledge and experience of an operator. The most often, operators within industrial conditions when preparing data to machining of the gears are making use of already defined conditions, enabling obtained accuracy class as assumed by design engineer of the gear.

With reference to adopted criterion of the optimization and obtained values of approaching angles of the grinding wheel, it has been ascertained that:

- for minimal coasting of the pass (gear wheel no. 2) it has been obtained respectively 7th accuracy class of manufacturing of the gear wheel,
- for minimal shift on measuring circle (gear wheel no. 4) it has been obtained respectively 5th accuracy class of manufacturing of the gear wheel,
- for minimal height of contact line on bearing surface of the tooth (gear wheel no. 1) it has been obtained respectively 5th accuracy class of manufacturing of the gear wheel,
- for preset value of the deflection angle (gear wheel no. 3) it has been obtained respectively 6th accuracy class of manufacturing of the gear wheel.

References

1. Oczoś, K.E., Porzycki, J.: Grinding. Basics and technology. WNT, Warsaw (1986)
2. Ochęduszko, K.: Gears. T. II. Execution and assembly. WNT, Warsaw (1992)
3. Oczoś, K.E., Marciniak, M.: Development of machine tool construction for grinding processes. Part 2. Mechanik **79**(3), 192–198 (2006)
4. PN-ISO 1328–1: 1997: Helical gears. Accuracy of implementation by ISO Definitions and values of deviations of the unilateral tooth sides
5. PN-ISO 1328-2: 1997: Definicje i wartości odchyłek pomiarowych złożonych i odchyłek bicia)
6. DIN 3962: 1978: Toleranze für Stirnradverzahnungen. Toleranze für abweichungen einzelner Bestimmungsgrößen
7. User manual for the Rapid 2000 shape-sensitive grinding machine for finishing the teeth of Höfler, Germany
8. www.klingelnberg.com/en/business-divisions/hoefler/cylindrical-gear-grinding-machines/detail-page/product/rapid-2000/. Accessed 20 Apr 2019
9. Płonka, S., Szadkowski, J., Matuszek, J., Kobiela, P.: Development of selected methods of machining gear teeth. Zeszyty Naukowe Politechniki Śląskiej. Seria: Transport. **87**(1929), 11–20 (2015)
10. Rosik, R., Świerczyński, J.: Influence of the MQL method on the parameters of shaping the active surface of the grinding wheel and the roughness of the surface layer of the workpiece. Inżynieria Maszyn **16**(1–2), 175–185 (2011)
11. Zyzak, P., Kobiela, P., Brożek, A., Gabryś, M.: The influence of the grinding strategy on the accuracy and roughness of gear teeth. Mechanik **91**(8–9), 737–740 (2018)
12. Płonka, S., Zyzak, P., Kobiela, P.: The effect of dressing the grinding wheel on the precision of the teeth grinded using a shaping method. Zeszyty Naukowe Politechniki Rzeszowskiej Mechanika t XXXIV Z **89**(4/17), 537–546 (2017)
13. Mehr, A., Yoders, S.: Efficient hard finishing of asymmetric tooth profiles and topological modifications by generating grinding. Gear Technol. J. Gear Manuf., 76–83 (2017)
14. Schalaster, R.: Fine grinding on Klingelnberg bevel gear grinding machines. Gear Technol. J. Gear Manuf., 30–32 (2019)
15. Burdick, R.: Parallel axis gear grinding: theory and application. Gear Technol. J. Gear Manuf., 77–81 (2000)
16. Gorgels, Ch., Schlattmeier, H., Klocke, F.: Optimization of the gear profile grinding process utilizing an analogy process. Gear Technol. J. Gear Manuf., 34–41 (2006)
17. Cannella, A.: Gear grinding today. Gear Technol. J. Gear Manuf., 22–27 (2017)
18. Strunk, S.: Surface structure shift for ground bevel gears. Gear Technol. J. Gear Manuf., 54–65 (2017)
19. Klocke, F., Schlattmeier, H.: Surface damage caused by gear profile grinding and its effects on flank load carrying capacity, Gear Technol. J. Gear Manuf., 44–53 (2004)
20. Podzharov, E.: Selection of the optimal parameters of the rack-tool to ensure the maximum gear tooth profile accuracy. Gear Technol. J. Gear Manuf., 34–37 (1999)
21. Richmond, D.: CNC gear grinding methods. Gear Technol. J. Gear Manuf., 43–50 (1997)
22. Türich, A.: Producing profile and lead modifications in threaded wheel and profile grinding. Gear Technol. J. Gear Manuf., 54–62 (2010)
23. Türich, A., Kobialka, C., Vucetic, D.: Innovative concepts for grinding wind power energy gears. Gear Technol. J. Gear Manuf., 39–44 (2009)

Evaluation of Geometrical Parameters of a Spur Gear Manufactured in an Incremental Process from GP1 Steel

Jadwiga Pisula[✉], Tomasz Dziubek, Łukasz Przeszłowski, and Grzegorz Budzik

Department of Mechanical Engineering, Rzeszow University of Technology, al. Powstańców Warszawy 8, 35-959 Rzeszów, Poland
jpisula@prz.edu.pl

Abstract. The main applicability criterion of an additive process is whether it allows a functional component to be directly manufactured. Obtaining the assumed dimensional and shape accuracy is an aspect of such functionality. In the case of additive manufacturing, geometrical accuracy depends on the additive technique, the use of which requires additional processes. This study presents the application of coordinate measurement methods to determine geometry reproduction accuracy for gears made in a hybrid process. The authors were able to determine the accuracy of gear geometry reproduction by means of Additive Manufacturing (AM) DMLS (Direct Metal Laser Sintering) techniques supplemented by subtractive finishing. Deformation for test models manufactured using the additive method were evaluated, along with the effect of post-processing, especially in terms of geometric deviations and the presence of shrinkage affecting the amount of allowance required in the correct subtractive manufacturing process. Model geometry was verified by contactless optical system ATOS II Triple Scan and a specialist coordinate measuring machine P40 by Klingelnberg. The analyses revealed that the machining allowance of 0.4 mm in the test model made it possible to obtain gear accuracy class 8 (in line with the assumptions) according to DIN 3962 when tooth spaces were machined using a non-specialist tool.

Keywords: Additive techniques · Subtractive manufacturing · Optical scanner · Coordinate measuring machine · Geometric accuracy · Spur gears

1 Introduction

The accuracy of gear prototypes fabricated in the process of Rapid Prototyping (RP) determined by means of coordinate measuring techniques offers quick and reliable verification of geometric errors [1, 2]. Gear accuracy class has a significant effect on the correct performance of a gear train and its durability [3]. Integration of Computer Aided Design (CAD) and Computer Aided Manufacturing (CAM/RP) systems as well as Coordinate Measuring Methods (CMM) enables significant acceleration of the production of gear train components with assumed manufacturing quality [4, 5].

© Springer Nature Switzerland AG 2020
G. M. Królczyk et al. (Eds.): IMM 2019, LNME, pp. 109–127, 2020.
https://doi.org/10.1007/978-3-030-49910-5_11

Considering the specifics of the accuracy of RP machines and the manufacturing tolerance assumed for components, it is necessary to combine additive and subtractive methods. However, a suitably designed hybrid manufacturing process requires the possibility of determining errors resulting from the additive process, and therefore the inclusion of adequate machining allowances. The thickness of the material layer removed in the subtractive process must allow for relevant restrictions, geometric errors arising from the accuracy of the RP machine, as well as any deformations of the physical model as a result of material shrinkage [1, 2, 4].

Test models were performed by Direct Metal Laser Sintering (DMLS). Most publications referring to additive laser techniques analyse mechanical and macroscopic properties as well as the microstructure of materials used in the AM process. For example, study [6] discussed the effect of material properties and conditions of the additive process on metallurgical mechanisms and resulting microstructural and mechanical properties of laser additive manufacturing of metallic components. Authors of [7] investigated the effect of the use of ultrasound during DMLS processing of stainless steel 316L on its mechanical properties and microstructure, whereas in [8] the impact of DMLS process parameters on macroscopic properties of aluminium alloy AlSi10Mg and the resulting microstructure was tested. Only in some publications do we find recommendations on the selection of manufacturing process parameters which define dimensional and geometric accuracy. In study [9], a surface roughness test of DMLS-manufactured aluminium samples was performed with varying process parameters such as laser power, scan speed or hatching distance. It was found that scan speed had the greatest impact on surface roughness. The authors of article [10] focused on benchmark artifacts for geometrical performance evaluation of additive manufacturing processes. An important issue which affects on dimensional and geometric accuracy as well as surface structure is the selection of component position on the build plate, component orientation in the printing space and the location of support structures. Publication [11] assessed the effect of surface orientation and the location of the component in the in printing space in the DMLS process on flatness errors (the orientation of the surfaces with respect to the platform and the position of the part in the build volume of the AM machine). Also, tests aimed at multi-criterion optimisation of the position of the component in the printing space in order to minimize the volumetric error along each building direction [12], to minimise the support structure volume, as well as support contact area [13] were carried out. Few research publications deal with hybrid manufacturing. Article [14] concentrates on research problems related to parameters of geometrical surface integrity after Fused Deposition Modelling (FDM) and turning by means of sintered carbide tools.

The present study is aimed at providing recommendations on hybrid manufacturing of a spur gear, the preliminary form of which was obtained by DMLS from GP1 powder. It also presents post-processing issues and determines the distribution and amount of subtractive machining allowances. A distinctive feature of the study is the visual presentation of changes in gear tooth flank surface topography after sintering, sandblasting and milling of tooth spaces using a non-specialised tool as well as the analysis of parameters defining the accuracy class of gears obtained in measurements on a WMP P40 instrument.

2   Research Methodology and Objectives

The subject of the analyses in the present study included model accuracy tests at different stages of the hybrid manufacturing process. Adequately prepared 3D-CAD models were used to fabricate gears with pre-formed teeth by means of DMLS (Direct Metal Laser Sintering) [15]. DMLS is a registered trademark of EOS GmbH based on the Laser Beam Melting (LBM) technology. The test models were made on an EOSINT M720 machine. The gears were made from Stainless Steel GP1 powder by EOS [16]; the material was high chromium stainless steel with chemical composition corresponding to steel No. 1.4542 according to EN 10027-2:2015 [17]. The additive manufacturing process was performed with the lowest possible layer thickness, i.e. 0.02 mm.

The test models were spur gears the geometry of which was defined by parameters presented in Table 1. The gears were made using the DMLS additive manufacturing technique, for which models included adequate allowances required during the subtractive phase (Fig. 1). Nine (9) gears were positioned on the work platform appropriately

Table 1.   Nominal test gear model parameters.

Designation	Description	Unit	Quantity
Number of teeth	z	–	19
Normal module	m_n	mm	3.0
Profile angle	α	deg	20
Helix angle	β	deg	0.0
Profile shift coefficient	x	–	0.0
Clearance	c	mm	0.75
Whole depth	h	mm	6.75
Face width	b	mm	10.0

Fig. 1. Test gear model fabrication using the DMLS additive technique directly after the sintering process with marked reference surfaces for bases.

to the machine layout. At consecutive manufacturing stages, gears described in Table 2 were used in the analysis of the results obtained during measurements.

Table 2. Test models.

Description	Designation	Image
DMLS gear after removal of support structures and preparation of base surfaces	KZB1	
Gear after sandblasting process	KZB2	
DMLS gear after teeth have been milled and other gear envelope surfaces have been turned	KZB3	

In order to determine the amount of shrinkage and make provisions for subtractive manufacturing, allowances of 0.1 mm, 0.2 mm, 0.3 mm and 0.4 mm were made on tooth space surfaces.

After the DMLS process the gears were subjected to post-processing, which involved cutting off the support structure. Base surfaces were also prepared in the form of the cylindrical surface of the hole and the face plane of the hub. In order to machine base surfaces, the gears were fixed in a three-jaw chuck gripping the cylindrical surface marked as A and supporting face B in Fig. 1. In addition, for gears in which tooth milling was foreseen, a groove was also cut in the base hole. A sample consisting of selected test models was sand-blasted with regular aloxite of grain size F80. The quartz flux was applied manually. A numerically-controlled lathe (Hass ST20Y) was used to make the gear envelope within a specific tolerance. Gear teeth were cut using a universal half-side mill with a diameter of 3 mm on the same machine, based on the assumption that the gear should be made on a non-specialist machine tool. The tooth machining method involved the tool moving along the tooth profile defined in the base CAD model.

3   Analysis of Dimension and Shape Accuracy

Two measuring machines were applied to verify gear manufacturing accuracy at each stage of fabrication. The first one was ATOS Triple Scan II by GOM [18], and the second other one was P40 by Klingelnberg, equipment dedicated directly to gear measurement [19]. The accuracy of the reproduction of geometry of gears made in the sintering process versus those manufactured using the subtractive process was determined on the basis of the measurements performed. Accuracy class of the product may be determined the basis of the values of parameters describing the geometry of gears recorded in reports from the coordinate measuring machine and a relevant standard. The effectiveness of the measuring process depends on the features and accuracy of the measuring instrument but also on the experience of the operator. Geometric features have a considerable influence on the results of gear measurements. The workflow of the measurement process must therefore be planned to include the accuracy with which the gear was manufactured, the type of the gear and the technique by which it was made, to account for measurement requirements. Selecting the correct measurement strategy allows us to shorten the time and maintain or even improve the accuracy of the measuring process [20, 21].

3.1   ATOS II Triple Scan – Making Measurements and Preparing Analysis

The aim of the study was to analyse the accuracy of the reproduction of a gear made by DMLS additive manufacturing as well as a gear after the subtractive process. Measurements were made with the help of a 3D Atos II Triple Scan optical scanner (Fig. 2).

Fig. 2. Measurements being taken with a 3D Atos II Triple Scan optical scanner.

Measurement data were processed in scanner's software environment ATOS Professional V7.5. The software is used for controlling the process of the acquisition of scanned geometry, processing measurement data and preparing reports on the accuracy of the reproduction of geometry of the analysed workpiece.

Considering measurement process automation capabilities, automatic measurement was applied on a rotary table forming part of the measurement system. Automation radically shortened digitization time, and an appropriate number of steps constituting the complete rotation of the table for a single measurement series was selected on the basis of a visual inspection of the geometry of the scanned model.

Global analyses of the entire gear surface were performed in order to determine the accuracy of the actual gear geometry. The results were presented as colour deviation maps (Figs. 3, 4, 5 and 6).

Fig. 3. Accuracy deviation maps for gear KZB1 manufactured with a specific tolerance zone generated in reference to the nominal model: a) 0.4 mm allowance; b) 0.3 mm allowance; c) 0.2 mm allowance; d) 0.1 mm allowance.

As the first step, the geometric accuracy of sintered gears was analysed after support structures were removed and base surfaces were prepared (Fig. 3). The reports indicate deviations in reference to the nominal CAD model, which did not include machining allowances, making it possible to determine the amount of shrinkage, and consequently to select the gear for subsequent machining. On the basis of the analyses it was decided that the gear should continue to be machined with an allowance of 0.4 mm (Fig. 3a).

In addition, to increase the precision of deviation distribution imaging, the gear was analysed in reference to the nominal model, which included the allowance on the toothed ring (Fig. 4).

Fig. 4. Accuracy deviation map for gear KZB1 manufactured with a specific tolerance zone generated in reference to the CAD model which includes a 0.4 mm allowance for machining.

Fig. 5. Accuracy deviation map for gear KZB2 manufactured with a specific tolerance zone generated in reference to the CAD model which includes a 0.4 mm allowance for machining.

The next stage of manufacturing, which involved sand-blasting selected test gears in which the outer metal surface had been removed from the teeth, was shown in the form of a global geometric analysis (Fig. 5). As in the analysis presented in Fig. 4, the comparison was made in reference to the nominal 3D-CAD model, the geometry of which had been expanded by machining allowance on the toothed ring (flanks, tooth space bottom except for tooth tip). Consequently, deviations in Fig. 5 are negative, but in reference to the nominal geometry of the final gear their values ensure required allowances for subtractive manufacturing.

The analyses suggest that a decrease in allowances during the phase in which the model is formed using additive techniques renders the remaining gears unsuitable for further subtractive processes due to material shrinkage, geometric errors in additive manufacturing and the application of sand-blasting.

Fig. 6. Accuracy deviation map for gear KZB3 manufactured with a specific tolerance zone generated in reference to the nominal CAD model.

After the final stage of the subtractive process by means of a slotting mill, the gear was assessed for accuracy in reference to the nominal 3D-CAD model (Fig. 6). The report indicates that the toothed ring is characterised by positive distribution of deviations.

Accuracy assessment by means of normal cross-sections at each manufacturing stage supplemented with numerical values at inspection points (Figs. 7, 8 and 9) was performed to obtain improved visualization of the distribution of deviations. The assessment helped determine precise values along the entire toothed ring in selected measurement points. Detailed analyses presented in Figs. 7, 8 and 9 confirm the correctness of the assumptions made on the basis of global deviation maps.

The results of gear geometry analysis are presented on test sample deviation maps. Because of the volume of the work only representative test results for 5 samples for each test model were presented. Note that the measuring method using the GOM optical system provides insight into model reproduction accuracy, in particular in reference to models manufactured using additive techniques. It also allows us to determine product geometry change trends relative to the nominal 3D-CAD model. Nevertheless, it should be emphasized that measurements using a coordinate measuring machine with software dedicated to gear accuracy analyses is highly recommended in the case of gears. The results of such measurements were supplied in the next section.

Fig. 7. Accuracy deviation map for gear KZB1 manufactured with a specific tolerance zone generated with deviations visualised in the required section in reference to the CAD model which includes a 0.4 mm allowance for machining.

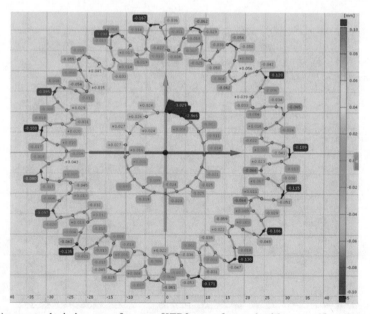

Fig. 8. Accuracy deviation map for gear KZB2 manufactured with a specific tolerance zone generated with deviations visualised in the required section.

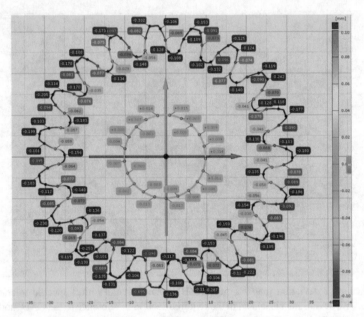

Fig. 9. Accuracy deviation map for gear KZB3 manufactured with a specific tolerance zone generated with deviations visualised in the required section.

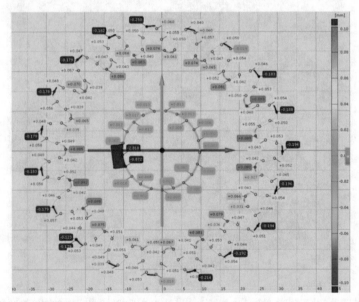

Fig. 10. KZB1 gear tooth flank topography chart.

3.2 Gear Measurements Using Klingelnberg P40 Together with Analysis

Test model measurements were made on a P40 coordinate measuring machine by Klingelnberg in laboratory conditions. The measurements were performed in accordance with spur gear measurement methodology on a coordinate measuring machine with the use of specialist software GINA [19]. The shortest possible spindle with a contact tip of diameter 1.5 mm was used, which ensured measurement error minimization. Measurement uncertainty on a P40 coordinate measuring machine according to VDI/VDE 2617 is determined on the basis of formula $UI = 18 + L/250$ [μm], where: L – numerical value of length measured in mm [19].

The measurement concerned spur gear teeth and was performed on each stage of the manufacturing process, which means that the gears were measured directly after fabrication on an EOS M270 machine, as well as after sand-blasting and tooth milling process.

Reports generated after gear measurements contained the following values and/or charts:

- profile deviations (total profile deviation $F\alpha$, profile form deviation $ff\alpha$, profile angle deviation $fH\alpha$),
- flank profile (flank line) deviations (total flank line deviation $F\beta$, flank line form deviation $ff\beta$, flank line angle deviation $fH\beta$),
- pitch deviations (individual pitch deviations fp, total pitch deviations Fp and maximum individual pitch deviation fpmax, largest pitch increment fumax, pitch fluctuations Rp, pitch span deviation for 8 consecutive measurements $Fpz/8$),
- gear's kinematic deviation (from the rotational movement) Fr,
- tooth thickness deviation Rs,
- tooth topography.

Gear measurement assessment was made on the basis of standard DIN 3962 [22]. A list of highest obtained values of parameters describing gear accuracy (listed above) together with corresponding accuracy class is provided in Tables 3, 4 and 5.

The values come from, respectively, a gear after sintering (KZB1), a gear after sandblasting (KZB2) and a gear after tooth milling (KZB3).

An accuracy class determined for at least one accurate parameter characterising gear teeth is, at the same time, a gear accuracy class. Therefore, based on the results of measurement obtained from all gears after the sintering process, presented in Table 3 for a representative gear of nominal geometry (excluding machining allowance), the gears are outside accuracy class 12. In addition, the amount of tooth thickness shrinkage was measured for the nominal sintered gear, for which maximum values ranges from 0.029 mm to 0.097 mm.

Table 3. Parameter values describing KZB1 gear tooth geometry (without allowance) after sintering together with accuracy class.

Parameters	Left flank		Right flank	
	Value obtained [μm]	Accuracy class	Value obtained [μm]	Accuracy class
Tooth profile parameters				
$f_{H\alpha}$	−62.7	12	−66.7	12
F_α	89.0	12	107	12
$f_{f\alpha}$	45.3	11	87.7	12
Flank line parameters				
$f_{H\beta}$	19.2	9	−12.4	8
F_β	79.5	12	72.0	12
$f_{f\beta}$	74.6	≫	68.9	≫
Pitch parameters				
f_{pmax}	34.5	11	30.6	10
f_{umax}	59.0	11	51.9	11
R_p	68	–	56.9	–
F_p	62.3	9	69.9	9
$F_{pz/8}$	43.9	9	43.5	9
Kinematic and tooth thickness parameters				
	Teeth concerned			
F_r	61.5		10	
R_s	86.9		12	

Tables 4 and 5 contain values of parameters characterising teeth of the gear which included a 0.4 mm machining allowance, respectively after the sand-blasting and tooth milling process. The sand-blasting process slightly decreases the values of parameters used for gear tooth geometry evaluation. This effect is also noticeable in the case of results for other gears from the sand-blasted sample; this is the reason why the gears are still outside accuracy class 12.

Machining gear teeth with a non-specialist tool (a half-side mill) following the path corresponding to gear tooth profile helps obtain a gear of accuracy class 8. Although tooth profile and flank line parameters were rated as accuracy class 7, final accuracy class was lowered due to pitch and tooth thickness deviations.

This study also presents the topography of a single gear tooth over consecutive manufacturing stages (Figs. 11 and 12). The topography was determined on 7 profiles distributed evenly over a certain profile assessment interval (interval $L\alpha$ defined by the standard) and 9 flank lines located within a relevant flank line assessment interval (interval $L\beta$ defined by the standard).

Table 4. Parameter values describing KZB2 gear tooth geometry (with 0.4 mm allowance) after sand-blasting and together with specified accuracy class.

Parameters	Left flank		Right flank	
	Value obtained [μm]	Accuracy class	Value obtained [μm]	Accuracy class
Tooth profile parameters				
$f_{H\alpha}$	−59.0	12	−62.9	12
F_α	82.2	12	66.4	11
$f_{f\alpha}$	63.0	12	63.9	12
Flank line parameters				
$f_{H\beta}$	37.5	11	−14.1	8
F_β	67.6	11	69.8	11
$f_{f\beta}$	63.7	≫	66.4	≫
Pitch parameters				
f_{pmax}	34.6	11	31.5	10
f_{umax}	47.2	11	47.9	11
R_p	65.6	–	58.4	–
F_p	61.2	9	78.4	10
$F_{pz/8}$	42.7	9	53.3	10
Kinematic and tooth thickness parameters				
	Teeth concerned			
F_r	94.3		11	
R_s	117.0		12	

Table 5. Parameter values describing KZB3 gear tooth geometry (with 0.4 mm allowance) after tooth milling and together with specified accuracy class.

Parameters	Left flank		Right flank	
	Value obtained [μm]	Accuracy class	Value obtained [μm]	Accuracy class
Tooth profile parameters				
$f_{H\alpha}$	6.4	7	7.3	7
F_α	6.7	5	8.7	6
$f_{f\alpha}$	3.4	4	5.1	5
Flank line parameters				
$f_{H\beta}$	10.3	7	−4.1	5

(continued)

Table 5. (*continued*)

Parameters	Left flank		Right flank	
	Value obtained [μm]	Accuracy class	Value obtained [μm]	Accuracy class
F_β	7.6	6	3.6	3
$f_{f\beta}$	6.9	7	2.8	3
Pitch parameters				
f_{pmax}	7.3	7	4.5	5
f_{umax}	12.8	8	6.8	6
R_p	12.8	–	8.4	–
F_p	10.4	4	19.5	5
$F_{pz/8}$	10.4	5	5.7	3
Kinematic and tooth thickness parameters				
	Teeth concerned			
F_r	19.8		6	
R_s	21.0		8	

On the basis of topography charts for sintered (Fig. 10) and sand-blasted gears (Fig. 11) it was determined that the sand-blasting process decreases elevation amplitude over profiles. Moreover, increased material depletion in the central section of the tip was reported, which was also observed in gear scans after the process (Fig. 5). The topography presented in Fig. 10 and 11 demonstrates also a change of tooth profile angle on both sides, which is further confirmed by negative values of profile angle deviations *fHα* (Table 3, Table 4). The topography of gears after tooth milling (Fig. 12) reveals a non-uniform distribution of remaining allowance, the amount of which increases from the tooth root towards the tip, obtaining the maximum value of 10 μm.

The optical measurement method enables the assessment of material shrinkage distribution after DMLS on every surface of the analysed component. The results reveal uneven material contraction on the surfaces of gear tooth spaces. The area near the tooth tip, where shrinkage is the greatest, is the critical location. On the basis of the foregoing, it was determined that a machining allowance of 0.4 mm is required. Measurements on a gear measuring centre revealed a post-DMLS tooth profile angle change (its value is greater than the anticipated value of profile angle). This finding should be taken into account when creating a model for additive manufacturing purposes.

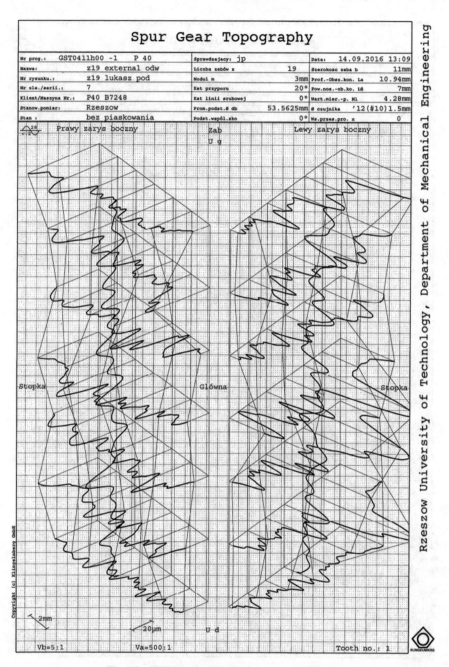

Fig. 11. KZB2 gear tooth flank topography chart.

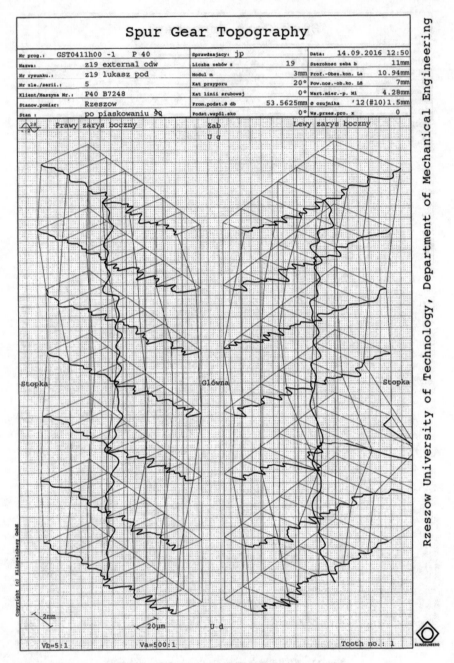

Fig. 12. KZB3 gear tooth flank topography chart.

4   Conclusions

The application of additive and subtractive manufacturing techniques to the analysed spur gear gave the expected results. This holds true for tooth milling by means of a universal tool – in this case a half-side mill (single piece production). The approach allows us to obtain gear accuracy class 8. We may expect that the application of specialist tools designed for gear tooth machining in envelope or form machining will result in an even higher accuracy class. Gear teeth must be checked after grinding, having previously been subject to heat treatment.

The process of making base surfaces is impeded due to the geometry of gripping surfaces obtained in the sintering process. Errors originating during the sintering process impact the location of base surfaces, and thus the size of designed allowances required to correctly implement the geometry of the toothed ring during subtractive machining.

Sand-blasting allows us to attain a much better geometrical structure of tooth surface, although it does not improve the gear's accuracy class. The process also fails to enhance the accuracy of the shape of base geometry, preventing the minimization of allowances for subtractive manufacturing.

Gear tooth measurements on a coordinate measuring machine after sintering and sand-blasting demonstrated that the profile angle for the right- and left-hand side of a tooth is greater than designed for the model (negative profile angle deviation). Furthermore, measurements of cylindrical surfaces on the gear envelope revealed their elliptical shape. It was also observed that the further the workpiece from the central position on the work platform, the greater the values of form deviations.

Tests performed by means of an optical measurement system led to similar results in terms of measurements and their analyses. This study acknowledges the usefulness of such machines in determining gear accuracy, as the results are presented in a manner convenient for interpretation in the case of additive techniques. However, due to the possibility of directly defining and determining gear accuracy class, the authors conclude that performing measurements on specialist contact measuring machines is relatively more convenient.

The tests performed in this study demonstrated that the designed allowances for subtractive manufacturing of 0.1–0.2 mm are too small in the view of the shrinkage which occurs during the sintering process. This is particularly noticeable in the tooth tip region.

When manufacturing gears with modules greater than those assumed in tests ($m_n = 3.0$ mm, see Table 1), recommendations concerning an allowance of 0.4 mm must be followed. For gears with a module lower or equal to 2.5 mm, the application of milling with a non-specialised tool is impossible due to tooth space width which includes allowance, and the milling cutter geometry, which is expected to ensure tool stiffness when machining the gear on a non-specialised machine tool.

The analyses presented here prove that ensuring the correct hybrid processing of gears requires the use of allowances for material shrinkage depending on tooth thickness on a specific diameter. It is also related to the effect of the deformations of base geometry, required for the correct performance of subtractive machining, an additional factor influencing the size of designed allowances.

References

1. Dziubek, T.: Application of coordination measuring methods for assessing the performance properties of polymer gears. Polimery **63**(1), 49–52 (2018). https://doi.org/10.14314/pol imery.2018.1.8
2. Pisula, J.: Geometric accuracy analysis of polymer spiral bevel gears carried out in a measurement system based on the Industry 4.0 structure. Polimery **64**(5), 353–360 (2019). https://doi.org/10.14314/polimery.2019.5.6
3. Lazarz, B., Wojnar, G., Czech, P.: Early fault detection of toothed gear in exploitation conditions. Eksploatacja i Niezawodnosc - Maint. Reliab. **1**, 68–77 (2011)
4. Pisula, J., Budzik, G., Przeszłowski, Ł.: An analysis of surface geometric structure and geometric accuracy of cylindrical gear teeth manufactured by the direct metal laser sintering (DMLS) method. Strojniški vestnik - J. Mech. Eng. **65**(2), 78–86 (2019). https://doi.org/10.5545/sv-jme.2018.5614
5. Rokicki, P., Kozik, B., Budzik, G., Dziubek, T., Bernaczek, J., Przeszłowski, L., Markowska, O., Sobolewski, B., Rzucidlo, A.: Manufacturing of aircraft engine transmission gear with SLS (DMLS) method. Aircr. Eng. Aerosp. Technol. Int. J. **88**(3), 397–403 (2016). https://doi.org/10.1108/AEAT-05-2015-0137
6. Gu, D.D., Meiners, W., Wissenbach, K., Poprawe, R.: Laser additive manufacturing of metallic components: materials, processes and mechanisms. Int. Mater. Rev. **57**(3), 133–164 (2012). https://doi.org/10.1179/1743280411Y.0000000014
7. Gale, J., Achuhan, A.: Application of ultrasonic peening during DMLS production of 316L stainless steel and its effect on material behaviour. Rapid Prototyp. J. **23**(6), 1185–1194 (2017). https://doi.org/10.1108/RPJ-09-2016-0140
8. Krishnan, M., Atzeni, E., Canali, R., Calignano, F., Manfredi, D., Ambrosio, E., Iuliano, L.: On the effect of process parameters on properties of AlSi10Mg parts produced by DMLS. Rapid Prototyp. J. **20**(6), 449–458 (2014). https://doi.org/10.1108/RPJ-03-2013-0028
9. Calignano, F., Manfredi, D., Ambrosio, E.P., Iuliano, L., Fino, P.: Influence of process parameters on surface roughness of aluminum parts produced by DMLS. Int. J. Adv. Manuf. Technol. **67**(9), 2743–2751 (2012). https://doi.org/10.1007/s00170-012-4688-9
10. Rebaioli, L., Fassi, I.: A review on benchmark artifacts for evaluating the geometrical performance of additive manufacturing processes. Int. J. Adv. Manuf. Technol. **93**(5–8), 2571–2598 (2017). https://doi.org/10.1007/s00170-017-0570-0
11. Rizzuti, L., De Napoli, S., Ventra, S.: The influence of build orientation on the flatness error in artifact produced by Direct Metal Laser Sintering (DMLS) process. In: Cavas-Martínez, F., Eynard, B., Fernández, Cañavate F., Fernández-Pacheco, D., Morer, P., Nigrelli, V. (eds.) Advances on Mechanics, Design Engineering and Manufacturing II, pp. 463–472. Springer, Cham (2019). https://doi.org/10.1007/978-3-030-12346-8_45
12. Li, Y., Zhang, J.: Multi-criteria GA-based Pareto optimization of building direction for rapid prototyping. Int. J. Adv. Manuf. Technol. **69**(5–8), 1819–1831 (2013)
13. Das, P., Mhapsekar, K., Chowdhury, S., Samant, R., Anand, S.: Selection of build orientation for optimal support structures and minimum part errors in additive manufacturing. Comput. Aided Des. Appl. **14**(S1), 1–13 (2017). https://doi.org/10.1080/16864360.2017.1308074
14. Krolczyk, G., Raos, P., Legutko, S.: Experimental analysis of surface roughness and surface texture of machined and fused deposition modelled parts. Tehničkivjesnik **21**(1), 217–221 (2014)
15. Kruth, J.P., Mercelis, P., Van Vaerenbergh, J., Froyen, L., Rombouts, M.: Binding mechanisms in selective laser sintering and selective laser melting. Rapid Prototyp. J. **11**(1), 26–36 (2005). https://doi.org/10.1108/13552540510573365
16. EOS GmbH Electro Optical Systems. http://www.eos.info/. Accessed 10 Aug 2018

17. EN 10027-2:2015, Designation systems for steels - Part 2: Numerical system
18. GOM: Precise Industrial 3D Metrology. https://www.gom.com. Accessed 29 June 2019
19. The Klingelnberg Group. https://www.klingelnberg.com. Accessed 10 Aug 2018
20. Goch, G.: Gear metrology. CIRP Ann. Manuf. Technol. **52**(2), 659–695 (2003)
21. Neumann, H.J.: Industrial Coordinate Metrology. VerlagModerneIndustrie, Lansberg, Lech (2000)
22. DIN 3962-1,2: Tolerances for cylindrical gear teeth. Part 1: Tolerances for deviations of individual parameters. Part 2: Tolerances for tooth trace deviation. Deutsche Normen (1978)

Technical and Economic Implications of the Combination of Machining and Additive Manufacturing in the Production of Metal Parts on the Example of a Disc Type Element

Dariusz Grzesiak[✉], Agnieszka Terelak-Tymczyna, Emilia Bachtiak-Radka, and Krzysztof Filipowicz

Faculty
of Mechanical Engineering and Mechatronics, West Pomeranian University of Technology Szczecin, Piastow Avenue 19, 70-310 Szczecin, Poland
dariusz.grzesiak@zut.edu.pl

Abstract. The article presents a detailed comparison of work inputs and costs necessary to implement two types of the manufacturing process of a typical machine part of a disc type. The subject of the comparison was a classical process - based solely on machining, and a hybrid process, in which a part mostly was formed in the additive manufacturing process. The results clearly show that in the unit production of metal parts, it is more cost-effective to use hybrid processes and maximize the use of additive manufacturing.

Keywords: Machining · Additive manufacturing · Selective Laser Melting · Hybrid manufacturing

1 Introduction

Additive manufacturing (AM) is currently one of the fastest developing technology [1, 2]. It is due to the incontestable advantages of the AM, such as no geometrical restrictions, no need to design and manufacture additional manufacturing equipment, no need to use consumable tools, ease of forming difficult-to-cut materials and many more [1, 6, 7]. The most popular for metallic parts manufacturing are the powder bed fusion technologies, in which successive thin layers of powder are selectively solidified by the energy source such as a laser or electron beam. The common feature of metallic powder based technologies and their significant disadvantage is low quality of produced parts which is the result of high surface roughness and low dimensional accuracy [8–10]. They result from the specificity of the process of selective melting of metal powder, in which the entire working space is filled with a powder of appropriate gradation (usually in a range of 40–50 μm). This powder at the interface between the part and the remaining space forms a rough surface with partially melted grains stuck and an irregular shape resulting from the liquefaction and quick cooling of the melted metal [4, 5].

G. M. Królczyk et al. (Eds.): IMM 2019, LNME, pp. 128–137, 2020.
https://doi.org/10.1007/978-3-030-49910-5_12

A strong development trend of the powder-based additive manufacturing technologies is their integration with the metal cutting technologies, such as milling or turning. Some manufacturers of machine tools, being aware of this direction of development, began to produce hybrid machine tools on which it is possible to combine additive manufacturing processes and machining [9, 10]. Nevertheless, it is currently a narrow niche, and such machines are extremely expensive. For this reason, the most popular method of implementing hybrid production processes is the use of separate devices. Considering that the time needed to form a certain unit volume of material is incomparably longer in the case of additive manufacturing, and in addition the cost of this material (in the powder form), is also much higher than the typical forms used in machining (rods, sheets), the use of a hybrid manufacturing process must be thought out and its costs very precisely defined.

This article presents research work aimed at a very accurate estimation of the costs of two methods of producing of a one piece of a typical metal element of the disc type – in a manufacturing process using only machining and in a manufacturing process combining additive manufacturing (Selective Laser Melting) and machining.

2 Materials and Methods

A representative part that was used to carry out the experiment was a disc element made of 17-4PH general purpose stainless steel. The geometrical shape of the element along with the required deviations in dimensions and shapes as well as the roughness are shown in Fig. 1. Two copies of the part were made using two different manufacturing processes, named "classic manufacturing process" and "hybrid manufacturing process". The initial assumptions for the conducted analyzes were omitting the costs of purchase and maintenance of machine tools and AM devices, and omitting the costs of purchasing technological equipment that did not wear during the experiment (tool holders, part holders, base plates for the SLM process). The classical manufacturing process was entirely carried out using a turning center CTX Ecoline with Y-axis positioning and driven tools. The manufacturing process scheme, as well as the NC programs, were prepared by a skilled technologist. The list of used cutting tools, together with purchase costs, is presented in the chapter "Results". The blank part was a rod with a diameter of 110 mm and a length of 3000 mm. In the classical manufacturing process, no 3D part model was needed to program the machine tool, so the time needed to prepare it was omitted here.

The hybrid process was carried out using the SLM Realizer II device and the above-mentioned machine tool. The first stage was the analysis of required dimensional and shape accuracy as well as surface roughness and classification of selected surfaces for additional machining. Next, a 3D CAD model was prepared with appropriate allowances on surfaces intended for post-processing, which were exported in STL format, which is a standard format for data exchange in incremental production. The STL file was used to generate a batch file needed to control the operation of the SLM device. Such a file contains the geometry of the part and supporting elements divided into successive layers and information about the work parameters (laser parameters, the beam moving strategy on the melted surface). The entire batch file preparation process was carried out using the device control software. In the next step the additive manufacturing process was carried out, which consisted of the actual work of the device lasting over 37 h (the operator was not required - the device is designed for autonomous operation), and preparation and post-process activities, which in the case of Selective Laser Melting are as follows: placing the base plate in the process chamber, loading the powder, heating the working platform, filling the process chamber with inert gas, removing the manufactured part from the process chamber and removing powder residues. The produced part was then cut off from the base plate (a band saw was used) and transported to a machining division, where a qualified operator selected cutting tools and technological parameters, and then machined selected surfaces, removing the remaining allowance.

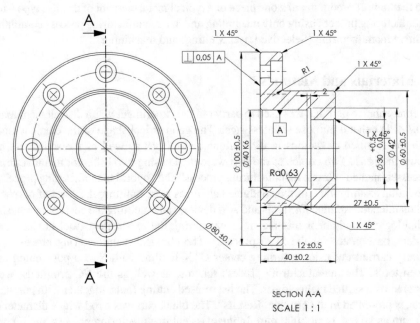

Fig. 1. The disc type part used for the analysis.

All operations related to the preparation and implementation of both manufacturing processes were considered in the analysis, as detailed in Table 2.

Measurements of dimensional and shape accuracy as well as roughness of manufactured parts were carried out using ZEISS Eclipse coordinate measuring machine No. 110283, HOMMEL TESTER T1000 roughness tester and electronic caliper. Measurement of electricity consumption during both processes was carried out using the Lumel ND 20 recorder. The current intensity was measured using transformers, whereas the voltage measurement was performed directly. The measurement was carried out on the main power supply of each device. A 200 A current intensity transformers with 5 A current intensity output accuracy class 3 were used for the measurement. The results were recorded using the PowerVIS recorder software. Electricity consumption was treated as a component of the total cost assuming a price per 1 kWh at the level of PLN 0.3877 and a price per 1kVarh at the level of PLN 0.5091. The assumed prices result from the Polish G11 tariff [11].

3   Results and Discussion

Analysis of the accuracy of manufacturing of both variants of the part showed that all dimensions and shapes are within the required tolerance fields. Surface roughness is also lower than the maximum allowed. The measurement results are shown in Table 1.

A summary of activities performed during the implementation of each of the processes and the time allocated to them are presented in Tables 2 and 3. The individual activities are grouped into stages containing tasks of a similar type. The stages named "Analytical activities" include actions related to the selection of tools, technological parameters or else the development of technological process diagrams. By assumption, these activities require the involvement of a highly qualified employee, e.g. a technologist. The stages named "Technical preparation" contain activities related to the preparation of technological machines for work. These activities require the involvement of a technical employee, such as a machine operator. The stages named "Manufacturing" contain activities directly related to the production of parts, such as fixing the item, machining, etc. In the case of a classic technological process, the technical employee is required to be involved in the whole production process, while in the case of a hybrid process, a technical worker is only needed in the part related to machining.

Figure 2a presents a graph showing the total duration of individual stages for both analyzed technological processes. It is evident that in the case of a hybrid process the most time-consuming stage is "Manufacturing", and the remaining stages are distinctly shorter than in the classic process. In the classic process, the "Analytical activities" stage took the most time, which is mainly related to the need to develop a technological process and the selection of tools and equipment. Also "Technical preparation" in this case took more time, mainly due to the necessity of equipping the machine with more tools than in the hybrid process.

Table 4 presents a list of costs incurred to manufacture parts in both variants of the technological process. In the case of the classical process, the purchase costs of used tools, electricity costs, material costs and labor costs of the technologist and operator were included in the calculation of total costs. In the case of the hybrid process, the total cost consisted of the cost of used for finishing cutting tools, the cost of the material (metal powder), the cost of electricity and the cost of the technologist and operator.

The cost of purchasing tools has been estimated based on catalog prices. For roughing, CCMT 120404EN-SM CTC2135 plates were selected, for finishing the DCMT 11T304EN-SM CTC2135 plate. The axial borehole was drilled with a folding drill bit XOMT 060304SN CTPP430. Drill holes were drilled with carbide drill bits WPC-VA.6,60.R.3D.IK.DIN6535.HA TIALN, and deepening in holes with countersink SE.N. 6.40X11.00.180°.DF.DIN373.

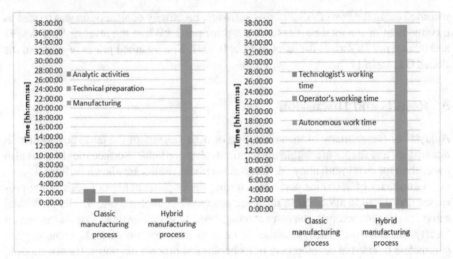

Fig. 2. A summary of the time needed for each type of activity (A) and the time consumed by each type of employee (B).

In the case of the classic process, the semi-finished product selected by the technologist was a rod with a diameter of 110 mm. The cost of purchasing the section of the rod needed for the experiment was 228.58 PLN. In the case of a hybrid process, the raw material was metal powder, of which 1 kg was priced by suppliers for an average of PLN 339.79. In the process, 1070 g of powder was used, which gave a real cost of PLN 363.57.

The cost of electricity consumption was estimated based on the actual consumption recorded during the implementation of technological processes, a summary of which is presented in Table 4. Due to the fact, that the SLM process lasted much longer, and during its lifetime the device was still working with typical parameters, clearly higher electricity consumption was observed in the case of the hybrid process. The total cost of electricity used to make parts in this variant was almost ten times higher than in the case of the classic process. Estimation of the cost of electricity, due to the absence of a reactive energy compensation system, considers both active and reactive energy consumption. In the case of using a compensation system, the cost of electricity can be reduced to PLN 22.23 for a hybrid process and up to PLN 1.67 for a classic process.

For the labor costs calculations, the report from the website Pracuj.pl [12] have been used. According to the report, the average earnings of the CNC machine operators reach PLN 4400 gross. That gives PLN 27.50 per hour assuming an 8-h working day. Comparatively, technologist's earnings amount to PLN 36.25 per hour. In hybrid process just like in the classic process, the technologist's working time selected for calculations was identical with the duration of "Analytical activities" while the operator's working time was identical with the duration of "Technical preparation" and "Manufacturing", but only in part related to machining, because the additive manufacturing process does not require the presence of staff.

Table 1. Results of the quality measurements.

Name of the geometric feature		Measured dimension [mm]	Nominal dimension [mm]	Upper tolerance [mm]	Lower tolerance [mm]	Measured deviation [mm]
Inner diameter D40mm	Classic MP	40.011	40	+0.003	−0.013	+ 0,011
	Hybryd MP	40.001				+ 0,001
pitch diameter D80mm	Classic MP	80.019	80	+0.1	−0.1	+0,019
	Hybryd MP	79.988				−0,012
Perpendicularity	Classic MP	0.015	0	+0.05	−0.5	+0,015
	Hybryd MP	0.066				+ 0,066
Outer diameter D100mm	Classic MP	100.03	100	+0.5	−0.5	+0.03
	Hybryd MP	99.75				−0.25
Thickness 12 mm	Classic MP	12.37	12	+0.5	−0.5	+0.37
	Hybryd MP	12.2				+0.2
Thickness 40 mm	Classic MP	40.11	40	+0.2	−0.2	+0.11
	Hybryd MP	39.94				−0.06
Outer diameter D60mm	Classic MP	59.95	60	+0.5	−0.5	−0.05
	Hybryd MP	59.93				−0.07
Inner diameter D30	Classic MP	30.09	30	+0.5	0	+0.09
	Hybryd MP	30.05				+0.05
Depth 27 mm	Classic MP	26.7	27	+0.5	−0.5	−0.3
	Hybryd MP	29.69				−0.31
Roughness Ra	Classic MP	0.63 μm	0.63 μm			0
	Hybryd MP	0.63 μm				0

The time needed for "Analytical activities" in the case of the hybrid process had included also the selection of cutting tools, technological parameters and preparation of machining programs. In practice, these activities would be to qualify for "Technical preparation". The machining has only consisted of the finishing transition on the front surface and cylindrical surface in one mounting, so the knowledge and skills of a qualified operator would be sufficient here. A summary of the engagement time of each employee in both technological processes is presented in Fig. 2b. Work time of machines in autonomous-mode for additive manufacturing is presented as a separate category because this possibility is an unquestionable advantage of this technology.

Table 2. A summary of activities performed during the implementation the classic manufacturing process.

Activity description	Stage in the process	Duration of activities [hh:mm:ss]	The duration of the stage
Classic manufacturing process			
Analysis of the part's technologicality	Analytic activities	0:15:00	2:53:00
Selection of a blank		0:03:00	
The division of the process into operations		0:20:00	
Selection of cutting tools and technological parameters		0:30:00	
Preparation of the NC programs		1:45:00	
Equipping the machine with tools	Technical preparation	0:55:00	1:25:00
Tool measurement		0:30:00	
Clamping the part	Manufacturing	0:00:30	1:03:05
Referencing		0:05:00	
Roughing of external surfaces		0:10:05	
Finishing of external surfaces			
Reclamping of the part with jaw replacement		0:30:00	
Referencing		0:05:00	
Drilling the axial hole		0:12:30	
Roughing of internal surfaces			
Finishing of internal surfaces			
Drilling holes for screws			
Counterboring the screw holes			
Together "Analytic activities"			2:53:00
Together "Technical preparation"			1:25:00
Together "Manufacturing"			1:03:05
Technologist's working time			2:53:00
Operator's working time			2:28:05
Autonomous work time			0:00:00
Together		**5:21:05**	

Table 3. A summary of activities performed during the implementation the hybrid manufacturing process.

Activity description	Stage in the process	Duration of activities [hh:mm:ss]	The duration of the stage
Hybrid manufacturing process			
CAD file preparation	Analytic activities	0:15:00	0:25:00
Batch file preparation		0:10:00	
Setting and fixing the base plate in the machine	Technical preparation	0:10:00	0:30:00
Filling the working chamber with inert gas and heating the base plate		0:20:00	
Manufacturing	Manufacturing	37:23:00	37:36:00
Removal of the part with the base plate from the machine and removal of powder residues		0:05:00	
Fixing the plate with the part on the band saw		0:05:00	
Cutting off the part from the base plate		0:03:00	
Selection of cutting tools and technological parameters	Analytic activities	0:10:00	0:25:00
Preparation of the NC programs		0:15:00	
Equipping the machine with tools	Technical preparation	0:25:00	0:35:00
Tool measurement		0:10:00	
Clamping the part	Manufacturing	0:00:30	0:07:35
Referencing		0:05:00	
Machining the face		0:02:05	
Finishing the inner diameter			
Together "Analytic activities"			0:50:00
Together "Technical preparation"			1:05:00
Together "Manufacturing"			37:43:35
Technologist's working time			0:50:00
Operator's working time			1:12:35
Autonomous work time			37:36:00
Together		**39:38:35**	

Table 4. A summary of costs.

Description of the cost	Value [PLN]
Classic manufacturing process	
Turning insert - 1 pcs.	71.41
Turning insert - 1 pcs.	57.03
Drill D6,6 - 1 pcs.	213.70
Drilling insert D36 - 2 pcs.	138.42
Counterbore drill - 1 szt.	67.02
Material	228.58
Electricity	4.06
The cost of the technologist's work	104.52
The cost of the operator's work	67.82
The total cost of the classical manufacturing process	**952.56**
Hybrid manufacturing process	
Turning insert - 1 pcs.	71.41
Turning insert - 1 pcs.	57.03
Material	363.57
Electricity	38.42
The cost of the technologist's work	30.09
The cost of the operator's work	33.28
The total cost of the hybrid manufacturing process	**593.79**

4   Conclusion

The calculations presented above clearly showed that according to the analyzed part, the hybrid process is proved to be more cost-efficient. This comparison assumes making only one workpiece. In the case of a larger number of manufactured workpieces, you should keep in mind a possibility of producing several parts simultaneously in a hybrid process, as well as spreading the cost of whole analytical work on the number of workpieces made. Also, the wear of cutting tools would change the final cost of producing each workpiece. According to the authors, the most important is the fact that additive manufacturing is a viable alternative to machining, and hybrid technological processes combining both technologies will soon become a standard in the production of metal parts.

References

1. Bingheng, L., Dichen, L., Xiaoyong, T.: Development trends in additive manufacturing and 3D printing. Engineering **1**(1), 85–89 (2015)
2. Dauderstädt, U., Askebjerb, P., Björnängen, P., Dürra, P., Friedrichs, M., Rudloff, D., Schmidt, J., Müller, M., Wagner, M.: Advances in SLM development for microlithography. Proc. SPIE – Int. Soc. Optical Eng. **7208**(720804), 1–7 (2009)
3. Yadroitsev, I., Bertrand, Ph, Smurov, I.: Parametric analysis of the selective laser melting process. Appl. Surface Sci. **19**(253), 8064–8069 (2007)
4. Dadbakhs, S., Mertens, R., Hao, L., Humbeeck, J., Kruth, J.: Selective laser melting to manufacture "In Situ" metal matrix composites: a review. Adv. Eng. Mater. **21**, 1–18 (2019)
5. Krutha, J., Froyen, B., Vaerenbergha, V., Mercelisa, P., Rombouts, M., Lauwers, B.: Selective laser melting of iron-based powder. J. Mater. Process. Technol. **1–3**(149), 616–622 (2004)
6. Grzesik, W., Niesłony, P., Habrat, W.: investigation of the tribological performance of AlTiN coated cutting tools in the machining of Ti6Al4V titanium alloy in terms of demanded tool life. eksploatacja i Niezawodnosc – Maintenance and Reliability, **21**(1), 153–158 (2019)
7. Matuszewski, M., Styp-Rekowski, M.: Polish machine-tools innovativeness. Determining factors. Obróbka Metalu **3**, 2–14 (2018)
8. Grobelny, P., Legutko, S., Habrat, W., Furmanski, L.: Investigations of surface topography of titanium alloy manufactured with the use of 3D print. Mater. Sci. Eng. **393**, 012108 (2018)
9. Królikowski, M., Krawczyk, M.: Obróbka skrawaniem oraz techniki przyrostowe jako integralne etapy procesu wytwarzania hybrydowego z metali w Przemyśle 4.0, Mechanik, **8–9**, 769–771 (2018)
10. Królikowski, M., Krawczyk, M.: Does Metal Additive Manufacturing in Industry 4.0 Reinforce the Role of Subtractive Machining?, Adv. Manuf. II, vol. 1 - Solutions for Industry 4.0, **1**, 150–164 (2019)
11. https://www.enea.pl/dlafirm/obsluga_klienta_i_kontakt/pliki_do_pobrania/wzory-umow-11.01.2019/umowa_kompleksowa_niekonsument-b-c/taryfa-dla-energii-elektrycznej-dla-grup-taryfowych-a-b-c-r.pdf
12. https://zarobki.pracuj.pl/stanowiska

Energy and Economic Indicators
of the Machining Process in Accordance
with the Requirements of ISO 50001

Monika Nowak and Agnieszka Terelak-Tymczyna(⊠)

West Pomeranian University of Technology, Szczecin, Poland
agnieszka.terelak@zut.edu.pl

Abstract. The paper presents terminology and applicable legal requirements in the field of energy efficiency. The proprietary methodology for energy management in manufacturing enterprises, taking into account environmental requirements as well as ISO 50001: 2018, has been presented. The purpose of the work is to increase the energy efficiency of production processes by optimizing machining processes. The authors have prepared a set of energy and economic indicators dedicated in particular to companies dealing with mechanical processing. Sample calculations were made for the turning process.

Keywords: Energy efficiency · Energy efficiency indicators · Energy management standard · Machining process

1 Introduction

Optimization of the decision-making process related to the reduction of electricity consumption and production costs has become one of the priority actions undertaken by enterprises in recent years. The reason for interest in businesses such actions are undoubtedly introduced legislation related to energy efficiency improvements.

The implementation of legal requirements relating to energy efficiency alike in EU countries [1, 2], as well as in many countries around the world [3, 4] has become an impulse to create an international standard for the Energy Management System (EnMS) - ISO 50001. The first attempt to harmonize the approach to energy management was the European standard EN 16001: 2009, which contained requirements for the electricity management system, allowing for systematic striving for continuous improvement of energy efficiency in enterprises, focusing on environmental issues. This standard was replaced in 2011 by the international standard ISO 50001 in which the emphasis was on energy consumption. The main concept presented in the standard is the concept of energy result, which takes into account aspects such as energy use, energy efficiency, energy intensity and energy consumption [5]. In 2018, an amendment to the standard was introduced that unifies the requirements in accordance with other ISO management standards. The standard introduces the concepts of standardization of energy efficiency indicators (EnPI) and related energy baselines (EnB).

© Springer Nature Switzerland AG 2020
G. M. Królczyk et al. (Eds.): IMM 2019, LNME, pp. 138–150, 2020.
https://doi.org/10.1007/978-3-030-49910-5_13

Restrictive tasks of governments around the world [1, 2] to reduce energy consumption have caused that interest in implementing the management system according to ISO 50001 standard is constantly growing. The number of valid certificates to ISO management system standards worldwide was shown on Fig. 1.

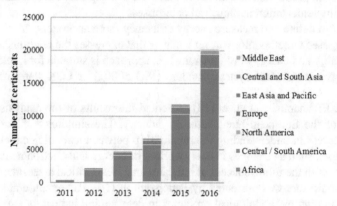

Fig. 1. The number of valid certificates to ISO management system standards worldwide [6].

In connection with the existing legal regulations and the constantly growing demand for electricity and other forms of energy, it became very important to make decisions to improve the efficiency of energy in all sectors of the economy, including the metal industry related to machining.

In the literature on the subject we find many items devoted to the analysis of energy efficiency in various industries.

Structure analysis, trends in energy consumption and energy efficiency indicators of business sectors occupied Al-Mansour in [7]. The document also provided an overview of energy efficiency policies and measures in Slovenia. Comparison of energy consumption in individual countries in the sectors of fossil energy, steel and cement production was done by Oda et al. in [4]. From the results obtained it can be concluded that regional differences in unit energy consumption are relatively large in the energy, steel and cement sectors. The results of the assessment of the potential for improving electrical performance in the Swedish pulp and paper industry are presented in [8] Blomberg et al. From those carried out by Blomberg et al. [8] analyses show that future energy efficiency programs could be better focused on clearly promoting technological progress. Whereas Hernández-Sancho et al. using energy efficiency indicators for environmental and economic analysis of sewage treatment plants located in Spain [9].

To support the decision-making process, it is necessary to develop indicators that allow to compare different technical solutions (including applied drives) and choose the most effective ones in terms of both energy and economic.

In recent years, a number of publications devoted to energy efficiency indicators and models which support decision-making process.

One such model is the standardized information model described by Bullinger et al. [10]. This model describes economic and environmental aspects, including energy throughout the product life cycle and provides this information for production planning and control. The important role of information in achieving energy efficiency shows Bunse et al. in [11]. Bunse et al. in [11] emphasizes that ICT (Information and Communication Technologies) tools and standardization are important elements enabling energy-efficient production and indicates that there is a gap between available solutions and actual implementation in industrial enterprises.

In order to define and measure energy efficiency more accurately, Giacone et al. in [12] he proposed a methodology using a matrix that expresses the relationship between inserted energy and energy of propulsion. This approach is suitable for implementation in an energy management system standard (ISO 50001) or LCA standard (e.g. ISO 14044).

In turn, Eichhammer et al. in [13] presented the results of the analysis of energy indicators of the European manufacturing industry. The simplicity of the indicators and the progress in understanding the relationship between technological changes and impact factors, such as changes in energy prices or energy and environmental policies, are key factors in the future success of indicator analyses. Critical assessment of energy efficiency indicators was presented by Pattersonn in [14]. Pattersonn emphasizes that the most common methodological problems in determining indicators are: the role of value assessments in the construction of energy efficiency indicators, the problem of energy quality, the problem of boundaries, the problem of joint production and the issue of isolation of the basic technical trend of energy efficiency from the aggregate indicator [14].

However, in contrast to the indicators contained in these publications, the methodology presented in this article has been adapted from the ISO 14031 standard and is oriented to the production processes of products requiring machining.

2 Methodology

The presented methodology for the application of energy score indicators for the production process of products using machining has been developed using the requirements of international standards ISO 14031: 2013 [15] and ISO 50001: 2018 [5, 16].

Energy efficiency in accordance with ISO 50001 is "an indicator or other quantitative relationship between results, services, goods or energy and energy input" [5].

Organizations, when planning the assessment of the energetic efficiency, should draw up in relation to the policy, objectives, energy tasks and other criteria regarding the energy efficiency of the organization. In addition, energy efficiency assessment can be helpful in:

1. identifying energy aspects,
2. determination of significant energy aspects,
3. setting criteria for energy efficiency,
4. assessing the effects of its energy activities with respect to predetermined criteria.

Energy efficiency assessment is an internal management process that uses indicators to compare information on the effects of past and present activity of the company in terms of energy use with the criteria of the effects of this activity. Indicators may apply to individual processes, a plant or a group of plants. Since the long-term development of the indicators is labor and time-consuming, it may be helpful to use the PDCA model (Fig. 2).

Fig. 2. PDCA model for energy efficiency assessment [5, 15, 17]

To define the EnPI indicators it is necessary:

1. define the system boundaries,
2. determine the parameters of inputs and outputs,
3. define the basis (energy baseline) for the assessment of indicators.

Table 1. Energy indicators classification [15–17].

	Indicators classification	Practical examples
1.	Absolute (total)	Monthly electricity consumption
2.	Relative (unit), in which data are most often referred to the size characterizing a given plant, such as the size of the plant, production in tones	Electricity consumption per unit of production
3.	Percentages that refer to the adopted standard or base value	Electricity consumption in the current year expressed as a percentage of electricity consumption in the reference year
4.	Aggregated, i.e. data or information of tAe same type, but from other sources, collected and expressed as a complex value	Total consumption of electricity from product production defined as the sum of electric energy consumption from individual processes/machines in which the product is created
5.	Weighted, i.e. data modified by applying their importance factor	

Data used as indicators or used to construct synthetic indicators may be expressed in absolute or relative terms and depending on the nature of the information and its intended use, they may be aggregated or weighted. The indicators classification is shown in Table 1.

The basic step in data analysis is to determine the functional unit to which we will collect all data. To avoid biased results, ISO 14031 recommends considering all relevant and reliable data that has been collected. Data analysis may include considerations regarding the quality, reliability, adequacy and completeness of the data necessary to develop reliable information [15, 16, 18].

Indicators determined as part of the energy efficiency assessment of processes in an enterprise can be divided into two basic groups, as shown in the Fig. 3.

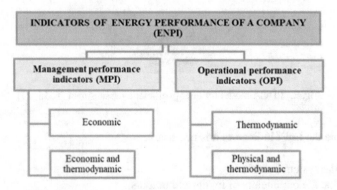

Fig. 3. Types of indicators in energy efficiency assessment.

In order to provide the basis for calculating the values of selected EnPI, data should be collected systematically from appropriate sources.

The organization of the data collection process affects their quality, which is why data should be collected and maintained in an orderly and systematic way. It should be ensured that the data is reliable, available, adequate, scientific and statistically reliable and verifiable. ISO standard 14031 recommends that data collection should be supported by quality control practices and quality assurance that determine compliance with the type and quality of data required to assess environmental performance and energy efficiency [15, 17].

Data analysis may include considerations regarding the quality, reliability, adequacy and completeness of the data necessary to develop reliable information [15, 17].

The EnPI results should be periodically and systematically reviewed in order to identify the possibilities of process improvement.

An overview of indicators and improvement of results may affect the improvement of data quality. These indicators are analyzed according to the following criteria:

1. Their analytical correctness and the possibility of evaluation,
2. The development of new or more useful energy performance indicators [15, 17, 19].

Based on the methodology presented in the ISO 14031 standard and by Putnam [16] and Jasch [18], indicators for the assessment of energy efficiency for the machining process have been developed. When developing the indicators, the critical assessment of indicators used by Pattersonn in [14] was also taken into account. In order to develop indicators for machining, a number of experimental studies were conducted regarding energy consumption during the machining process. These studies were conducted in the years 2016–2019 both in the enterprise from the metal industry and in the Technology Hall of the West Pomeranian University of Technology in Szczecin. A portable measuring device consisting of a Lumel recorder, current transformers, data converter and portable computer was used for the tests. Data recording took place on the main power supply of the machine. The following parameters were recorded: current, voltage, active power, reactive power, apparent power, active energy, reactive energy, power factor and THD harmonics. Measurements were made for mass-produced products. Based on the collected data, a set of indicators was developed, presented in the next chapter.

3   Energy and Economic Indicators of the Machining Process

Planning result of the energy ratios (EnPI) for machining has to be taken into consideration [5, 15, 20]:

1. A full range of operations and treatments performed during the machining process,
2. energy policy,
3. present processes in the form of block diagrams with specific input and output parameters,
4. within the framework of separate partial processes, determine the size of the inputs and outputs,
5. determine the basis (energy baseline) for the assessment of indicators,
6. information necessary to meet legal requirements,

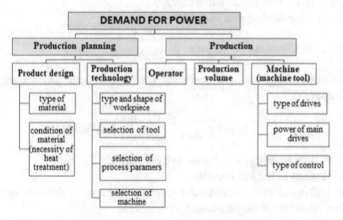

Fig. 4. Elements affecting the power demand of the machine tool.

7. costs and profits related to energy efficiency,
8. information on the best available BAT techniques.

For the machining process, factors influencing the power demand in the process were divided into two categories (Fig. 4).

Taking into account the above considerations, the following EnPI indicators (1) and (6) can be determined for the machining process:

1. Management Performance Indicators:

   a. Unit cost of electricity J_{EnC} (1)

   $$J_{EnC} = \frac{\sum(En_{in} \cdot Cj)}{Pr}, PLN/pcs \qquad (1)$$

   where: En_{in} – the amount of energy delivered to the process, kWh (for active energy) or kVARh(for reactive energy)
   Pr – production volume expressed in pcs
   c_j – unit price of energy in PLN/kWh

   b. The total cost of energy consumption per batch of ordered products (2)

   $$EnC = \sum(En_{in} \cdot c_j) \cdot P_{pr}, PLN/batch \qquad (2)$$

   where: En_{in} – the amount of energy delivered to the process, kWh (for active energy) or kVARh (for reactive energy)
   c_j – unit price of energy in PLN/kWh
   P_{pr} – the size of the production batch.

   c. Daily cost of electricity usage (3)

   $$D_{EnC} = D_{En} \cdot c_j, PLN \qquad (3)$$

   where: D_{En} – daily amount of energy delivered to the process, kWh (for active energy) or kVARh (for reactive energy)
   c_j– unit price of energy in PLN/kWh.

2. Operational Performance Indicators:

   a. Individual energy demand delivered per unit of J_{PW} product expressed in kWh/pcs (4)

   $$J_{PW} = \sum_{i=1}^{n}\left(\frac{P_{ei} \cdot t_{ji}}{3600000}\right), kWh/pcs \qquad (4)$$

   where: P_{ei} - cutting power for the i-th operation
   t_{ji} – duration of the i-th operation.

   b. The individual energy demand necessary to perform the i-th technological operation JOTi for single product kWh/pcs (5)

   $$J_{OTi} = \frac{P_{ei} \cdot t_{ji}}{3600000}, kWh/pcs \qquad (5)$$

c. Total energy demand per batch of ordered products (6)

$$En = \sum En_{in} \cdot P_{pr}, \, kWh \, or \, kVARh \qquad (6)$$

where: En_{in} – the amount of energy delivered to the process, kWh (for active energy) or kVARh(for reactive energy)

P_{pr} – the size of the production batch expressed.

3.1 Energy and Economic Indicators of the Rolling Process

The methodology presented in the previous chapters was used to determine the energy efficiency of shield and roller production. The energy and economic indicators of the rolling process were calculated for the production on two different machines.

In the first case, machining was carried out on the lathe DMG CTX 520 Linear CNC. The order quantity from the customer was 200 shield type items. The processing time of one element is 4 min 15 s. The plant works on one shift with a half-hour break for the employee. The analysis does not take into account the reactive power compensation by the manufacturing plant. The basic characteristic data of the machining process together with prices for energy are presented in Table 2.

In the second case, the lathe DMG CTX 310 Ecoline with synchronous drive was used for production. Machining was carried out for similar conditions. This time order quantity from the customer was 240 roller type items. The processing time of one element is 3 min 38 s. The plant works on one shift with a half-hour break for the employee. The analysis does not take into account the reactive power compensation by the manufacturing plant. The basic characteristic data of the machining process are presented in Table 3. Prices for energy for both lathes were the same.

Table 2. Characteristics of technical and economic data for the production of a shield type on a machine with linear drives

Lp.	Parameters description	Value	Unit
1	Set-up time	1414	sec
2	Processing time of one element	255	sec
3	Price for active electricity, c_j	0,5091	PLN/kWh
4	Price for inductive reactive electricity, c_j	0,5091	PLN/kWh
5	The size of the production batch, P_{pr}	200	pcs
6	Amount of active energy consumed per one element, En_{in}	0,4292	kWh
7	Amount of reactive capacitive energy consumed per one element, En_{in}	0	kVarh
8	Amount of reactive inductive energy consumed per element, En_{in}	0,4282	kVarh
9	tg φ	1,52	
10	Amount of active energy consumed during set-up time, En_{in}	1,6395	kWh
11	Amount of reactive inductive energy consumed during set-up time, En_{in}	2,9155	kVarh
12	Daily production time,	27000	sec
13	Daily production volume	100	pcs

The daily amount of energy delivered to the process D_{En} were calculated as:

– for active power (7)

$$D_{En} = \left(\frac{\text{Amount of active energy consumed per one element}}{\text{Daily production volume}} \right) +$$

$$\text{Amount of active energy consumed during setup time} = 0,4292 \cdot 100 + 1,6395 = 44,56\,kWh \quad (7)$$

– for reactive power (8)

$$D_{En} = \left(\frac{\text{Amount of reactive inductive energy consumed per one element}}{\text{Daily production volume}} \right) +$$

$$\left(\frac{\text{Amount of reactive capacitive energy consumed per one element}}{\text{Daily production volume}} \right) +$$

$$\text{Amount of reactive inductive energy consumed during setup time}$$
$$= (0,4282 \cdot 100) + (0 \cdot 100) + 2,9155 = 45,73\,kVARh \quad (8)$$

The amount of energy delivered to the process, $En_{in,}$ were calculated as:
– for active power (9)

$$En_{in} = \left(\frac{\text{Amount ofactive energy consumed per one element}}{\text{The size ofthe production batch}} \right) +$$

$$\text{Amount ofactive energy consumed during setup time} =$$
$$0,4292 \cdot 200 + 1,6395 = 87,48\,kWh \quad (9)$$

Table 3. Characteristics of technical and economic data for the production of a roller type on a machine with synchronous motors

Lp.	Parameters description	Value	Unit
1	Set-up time	405	sec
2	Processing time of one element	218	sec
3	The size of the production batch, P_{pr}	240	pcs
4	Amount of active energy consumed per one element, En_{in}	0,1448	kWh
5	Amount of reactive capacitive energy consumed per one element, En_{in}	0	kVarh
6	Amount of reactive inductive energy consumed per element, En_{in}	0,1782	kVarh
7	tg ф	1,23	
8	Amount of active energy consumed during set-up time, En_{in}	0,1297	kWh
9	Amount of inductive reactive energy consumed during set-up time, En_{in}	0,1380	kVarh
10	Daily production time	27000	sec
11	Daily production volume	120	pcs

Table 4. Energy and economic indicators for rolling process.

Energy and economic indicator		Value		Unit
		DMG CTX 520 Linear CNC	*DMG CTX 310 Ecoline*	
Management Performance Indicators	The unit cost of electricity JEnC	0,3919	0,1463	PLN/pcs
	The total cost of energy consumption per batch of ordered products EnC	78,39	35,12	PLN/batch
	The daily cost of electricity consumption DEnC	39,19	17,56	PLN
Operational Performance Indicators	The unit demand for active energy supplied per unit of produced product JPW	0,4374	0,1453	kWh/pcs
	The unit demand for reactive energy supplied per unit of produced product JPW	0,4428	0,1787	kVARh/pcs
	The unit energy demand necessary for the execution of i-th technological operation JOTi per product	0,8574	0,3230	kWh/pcs
	Total demand for active energy per batch of ordered products	87,48	34,88	kWh/batch
	Total demand for reactive energy per batch of ordered products	88,56	42,90	kVARh/batch

– for reactive power (10)

$$
\text{En}_{in} = \left(\begin{array}{c} \text{Amount of reactive inductive energy consumed per one element} \cdot \\ \text{The size of the production batch} \end{array} \right) +
$$

$$
\left(\begin{array}{c} \text{Amount of reactive capacitive energy consumed per one element} \cdot \\ \text{The size of the production batch} \end{array} \right) +
$$

$$
\text{Amount of reactive energy consumed during setup time} =
$$

$$
(0,4282 \cdot 200) + (0 \cdot 200) + 2,9155 = 88,56\, kVARh
$$

(10)

The data presented in the Table 2 and Table 3 were used to determine the EnPI indicators. These indicators were calculated for the rolling process of a workpiece shield type using the formulas (1) and (6). Result of calculation present in Table 4 and on Fig. 5 and 6.

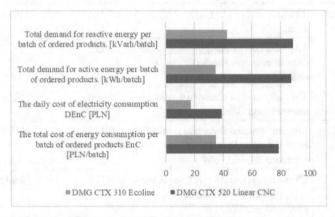

Fig. 5. Total energy and economic indicators for rolling process.

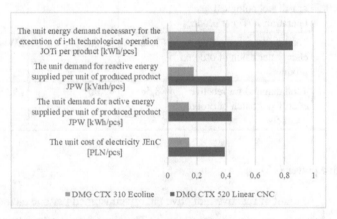

Fig. 6. The unit energy and economic indicators for rolling process.

From the obtained results, it can be concluded that the daily demand for electricity increases the production costs by 39.19 PLN for a lathe with linear drivers and by 17, 56 PLN for a lathe with synchronous motors. The total electricity demand is 87.48 kWh and 88.56 kVARh of inductive reactive electricity for a lathe with linear drivers and 34, 88 kWh and 42, 90 kVARh for a lathe with synchronous motors.

To increase energy efficiency, an inductive reactive power compensator should be installed first. This action could reduce the energy intensity of analyzed production by 88.56 kVARh and reduce the cost of energy demand by 57% for a lathe with linear drivers and analogously 42, 90 kVARh and 62% for a lathe with synchronous motors.

As it could be seen in the Figs. 4 and 5 the energy demand for lathe with synchronous motors are most twice smaller than for a lathe with linear drivers.

4 Summary

In connection with the existing legal regulations and the constantly growing demand for electricity and other forms of energy, it became very important to make decisions to improve the efficiency of energy in all sectors of the economy, including the metal industry.

The presented methodology has been developed with manufacturing companies in mind and allows searching for energy waste in machining processes. Importantly, the methodology has been developed based on applicable legal regulations and in accordance with the concept of continuous improvement.

The paper indicates a number of aspects that may affect energy consumption during machining processes. During the research, the focus was on the means of work used in the process, i.e. machine tools and more accurately used drives.

The comparison of energy demand for lathe with synchronous motors are most twice smaller than for a lathe with linear drivers. In this case the energy efficiency can be improved by using an inductive reactive power compensator. In modern machine tools for machining, motors and systems that reduce the energy consumption of the machining process are already assembled.

In the future, it would be necessary to include in the proposed indicators the share of the remaining parameters of the machining process, in particular the individual drive units of the machine tool and the production planning method itself.

References

1. European Commission.: Directive 2012/27/Eu Of The European Parliament and of the Councilof 25 October 2012on energy efficiency, amending Directives 2009/125/EC and 2010/30/EU and repealing Directives 2004/8/EC and 2006/32/EC (2012). http://eur-lex.eur opa.eu
2. European Commission: Directive (EU) 2018/2002 of the European Parliament and of the Council of 11 December 2018 amending Directive 2012/27/EU on energy efficiency (2018). http://eur-lex.europa.eu
3. Lan-Bing, L., Jin-Li, H.: Ecological total-factor energy efficiency of regions in China. Energy Pol. **46**, 216–224 (2012)
4. Oda, J., Akimoto, K., Tomoda, T., Nagashima, M., Wadaa, K., Sano, F.: International comparisons of energy efficiency in power steel and cement industries. Energy Policy **44**, 118–129 (2012)
5. Standard ISO 50001:2018: Energy Management System (2018). (The International Organization for Standardization)
6. Charlet, L.: The ISO Survey (2018) (The International Organization for Standardization). https://www.iso.org/the-iso-survey.html
7. Al-Mansour, F.: Energy efficiency trends and policy in Slovenia. Energy **36**(4), 1868–1877 (2011)
8. Blomberg, J., Henriksson, E., Lundmark, R.: Energy efficiency and policy in Swedish pulp and paper mills: a data envelopment analysis approach. Energy Policy **42**, 569–579 (2012)

9. Hernández-Sancho, F., Molinos-Senante, M., Sala-Garrido, R.: Energy efficiency in Spanish wastewater treatment plants: a non-radial DEA approach. Sci. Total Environ. **409**(14), 2693–2699 (2011)
10. Bullinger, H.J., Steinaecker, J.V., Weller, A.: Concepts and methods for a production integrated environmental protection. Int. J. Prod. Econ. **60–61**, 35–42 (1999)
11. Bunse, K., Vodicka, M., Schönsleben, P., Brülhart, M., Ernst, F.O.: Integrating energy efficiency performance in production management - gap analysis between industrial needs and scientific literature. J. Cleaner Prod. **19**, 667–679 (2011)
12. Giacone, E., Manco, S.: Energy efficiency measurement in industrial processes. Energy **38**, 331–345 (2012)
13. Eichhammer, W., Mannsbart, W.: Industrial energy efficiency. Indicators for a European cross-country comparison of energy efficiency in the manufacturing industry. Energy Policy, **25**(7–9), 759–772 (1997)
14. Pattersonn, M.G.: What is energy efficiency? Concepts, indicators and methodological issues, Energy Policy, **24**(5), 377–390 (1996)
15. Standard PN-EN ISO 14031:2013 Environmental management - Environmental Performance Evaluation – Guidelines (2013). (Polski Komitet Normalizacyjny)
16. Putnam, D.: ISO 14031: Environmental Performance Evaluation, Draft Submitted to Confederation of Indian Industry for publication in their Journal, September 2002
17. Sobczak, T., Terelak, A.: Wykorzystanie wskaźników do oceny efektów działalności środowiskowej -ISO 14031 cz.1. Problemy jakości, vol. 38, pp. 39–42 (2006). (in Polish)
18. Jasch, Ch.: Environmental performance evaluation and indicators. J. Cleaner Prod. **8**(1), 79–88 (2000)
19. Terelak-Tymczyna, A., Pawlukowicz, P.: Wskaźniki wyniku energetycznego dla procesu obróbki skrawaniem na przykładzie toczenia, Mechanik, vol. 87 (2014) (in polish)
20. European Commission, Reference Document on Best Available Techniques for Energy Efficiency. Sevilla:EC JRC (2009). http://eippcb.jrc.es/reference/BREF/ENE_Adopted_02-2009.pdf

Analysis of Cutting Force and Power Under the Conditions of Minimized Cooling in the Process of Turning AISI-1045 Steel with the Use of the Parameter Space Investigation Method

Radoslaw W. Maruda[1]($\boxtimes$), Stanislaw Legutko[2], Roland Mrugalski[3],
Daniel Dębowski[1], and Szymon Wojciechowski[2]

[1] Faculty of Mechanical Engineering, University of Zielona Gora, 4 Prof. Z. Szafrana Street,
65-516 Zielona Gora, Poland
r.maruda@ibem.uz.zgora.pl
[2] Faculty of Mechanical Engineering and Management, Poznan University of Technology,
3 Piotrowo Street, 60-965 Poznan, Poland
[3] Volkswagen Motor Polska sp. z o.o., 1 Strefowa Street, 59–101 Polkowice, Poland

Abstract. The work presents methods of cooling during turning of AISI-1045 steel: dry machining, minimum quantity cooling (MQC) and minimum quantity cooling with EP/AW (MQC + EP/AW) additives. Due to the increased number of variables (change of the parameters of cutting and forming emulsion mist), the method of Parameter Space Investigation was employed. The measurement of the cutting force was performed by means of a Kistler dynamometer type 9129AA and power consumption was measured with the use of a network parameter measurement device, MPS7. It has been found in the investigation that the air flow intensity has a crucial impact on the formation of the active medium in the MQC method. The lowest values of the cutting force and power have been observed when using the MQC + EP/AW method. Addition of phosphorate ester to the emulsion mist has resulted in an improvement of lubricating properties of the active medium by direct influence of the chemically active compounds in the cutting zone and, consequently, significant reduction of the cutting force and power in the process of turning the AISI-1045 steel.

Keywords: EP/AW additives · Cutting force · Cutting power · MQC · Dry cutting

1 Introduction

Due to the growth of society awareness related to ecological aspects of manufacturing, the method of cooling based on minimal quantities of cooling and lubrication has become an interest of many scientists [1–4]. As result of the application of very small quantities of CCS in the MQL or MQC methods, both the tool, chip, the object being machined and

© Springer Nature Switzerland AG 2020
G. M. Królczyk et al. (Eds.): IMM 2019, LNME, pp. 151–162, 2020.
https://doi.org/10.1007/978-3-030-49910-5_14

the machine working space remain almost completely dry. In the MQL method, the main lubricating agent is oil [5, 6] and, in the MQC method, it is emulsion in which the main agent is water constituting, in most cases 92% of the emulsion composition [7, 8]. The methods in which the basis is, additionally, lubricating agents or cooled air is provided to machining zone are called the MQCL (minimum quantity cooling lubrication) method [9, 10]. Small sizes of the droplets result in increase of the cutting zone penetration [11, 12] and, consequently, are more advantageous, especially for applications in the micromachining processes with the relatively small chip thicknesses. Moreover, the distribution and diameters of the droplets are related to the area of wettability and can have significant influence on the cutting process, as well as on other important features, such as thermal and chemical stability of the cooling and lubricating agent [13].

Reduction of the friction force in the contact zone of the cutting edge-material under machining-chip reduces the quantity of heat generated between the tool and the chip. Small droplets vaporize quicker from the cutting zone and, consequently, take more heat [10], which also contributes to an increase of the tool cooling. However, the effectiveness of lubricating and cooling depends mainly on the intensity of the air flow [11, 12, 14], intensity of the flow of the cooling and lubricating medium [14, 15], the nozzle positioning in relation to the cutting zone [14], the angle of the channels inclination in the tool [16, 17] and on the kind of the cooling and lubricating fluid used during machining [18].

For assessment of the cutting process in machining metals, it is also necessary to determine the force and power of cutting by analysing the kinematic – technological aspects and the physical mechanism of the process, such as: properties of the material being machined, the cutting edge geometry, kind of technological operation and the direction of the cutting motion. Both values are necessary for determination of the machine power, tool durability and for determination the stability and dynamics of the O-U-P-N system. The power of cutting is mainly determined by the cutting speed and cutting force, F_c, being a projection of the total force, F, on the direction of the main motion [19]. The investigation results shown in literature concerning limitation of the use of CCS or its complete elimination on the cutting force do not indicate the application trend for the individual methods of cooling and lubrication [20, 21].

Behera et al. [22] have proposed a new model for determining the cutting force. This approach proposes a novel theory of modified two contact zones (sticking-sliding) instead of the theory of one contact zone which models the relationship in the tool-chip interface. Additionally, the authors have elaborated a mechanical model which initially determines the friction coefficient in the sliding contact, variable in terms of the MQL parameters and the cutting parameters. That model, however, does not assume the diameter and number of the droplets supplied to the cutting zone during application of the MQL method.

Mia et al. [1] have used Taguchi method to determine the force values and roughness of the machined surface of hardened AISI 4140 steel after the process of milling depending on the variable cutting parameters and the intensity of the cooling and lubricating medium flow. The application of this method of investigation planning has allowed the authors to find that the cutting force is most influenced by the travel and least by the intensity of the mass flow of the active medium. The largest influence on the surface roughness is that of the flow intensity of the machining medium.

Park et al. [23] have combined the MQL method with the cryogenic method in milling of a titanium alloy and compared it to dry and wet machining in respect of the cutting force and tool wear. The authors have proved that the combination of the MQL method and cryogenic cooling is most promising due to the increased effectiveness of cooling and lubricating the cutting zone.

Literature is short of works presenting cooling methods based on emulsion, particularly works concerning change of the parameters of cutting and those of forming the emulsion mist (MQC method). That is why the purpose of experimental tests is to determine the influence of the parameters of forming the active medium and the parameters of cutting on the cutting force and cutting power as compared to dry machining.

2 Experimental Procedure

The tests were performed on a universal centre lathe, CU 502. The tool was a lathe tool with a CSRNR2525 holder and a SNUN1204008-PF plate. The material of the cutting plate was a sintered carbide with TiAlN coating with the thickness of about 3 μm. The tests have been performed in a wide range of cutting speeds, from 70 to 420 m/min and with the travel of 0.05 to 0.3 mm/rev and with constant cutting depth of 0.4 mm. The material under machining was carbon steel with the chemical composition: C − 0.42–0.5%; $Si_{max} - 0.4\%$; $S_{max} - 0.045\%$; $P_{max} - 0.04\%$; $Mn_{max} - 0.5–0.8\%$; $Cr_{max} - 0.3\%$; $Cu_{max} - 0.3\%$; $Ni_{max} - 0.4\%$ and with the hardness of 250 HB.

In MQC technique, the active medium was formed with application of a Micronizer device which enables the control of the flow intensity of medium and the air flow intensity. The working pressure of the compressor was 0.48 MPa. The range and values of the emulsion mist parameters according to the PSI method can be found in Table 1.

In the MQC method, the active medium was the emulsion concentrate on the base of mineral oil without additional lubricating agents and nitrates, chlorine, formaldehydes. The concentration applied was 7%, according to manufacturer's recommendation for turning the AISI-1045 steel. As the EP/AW additive, esters based on calcium and phosphorus with the concentration of 5% (as recommended by the producer) have been used.

The cutting forces were measure with a Kistler dynamometer type 9129AA. This is a four-channel piezoelectric dynamometer for measurement of the three cutting forces: the main one F_c; the feed one F_f and the thrust one F_p. The measurement signal from the dynamometer is transformed into voltage for the individual channels and transferred to the multi-channel amplifier type 5070A and then transferred to the data acquisition system. The frequency of the signal sampling has been set to 600 Hz in accordance with the producer's recommendation. The total force F, has been calculated from the formula:

$$F = \sqrt{F_c^2 \cdot F_f^2 \cdot F_p^2} \qquad (1)$$

The power consumption measurement has been performed with the use of a network parameter measurement gauge MPS7. The MPS7 gauge is intended for measurement and indication of three-phase energetic network. The signal taken from the circuit of the energetic network is realized in the device by the method of sampling. The results are sent to a computer through an RS-232C interface and the collected samples of signal courses allow the basic measured magnitudes to be determined.

Because of the large number of variables during the experiment, the Parameter Space Investigation method (PSI) was employed [24, 25]. The method allows for the determination of the function $(Y = f(X_i))$ in a multi-dimensional space the values of which can be calculated in selected points. Let us assume that we want to obtain certain pieces of information on the function behaviour in the whole multi-dimensional space or in any area of it. For a single variable, there is no problem with its uniform distribution which had been obtained by the division of the variables' range into ñ equal elements and localization of the point, where a sequence of n points (network) is evenly distributed in the space. Nevertheless for the multiple variables, the notion of homogeneity is ambiguous. In case of the each of variables one can divide in the same way as for a single variable, However for a number of variables equal to N, one can obtain N^m points (cubical network). On the other hand, the idea of homogeneity should be not dependent from the number of points and, what is more, application of networks containing many points influentially hinders the solution of practical problems. For that reason, in the PSI method, the points of experimental tests are situated in the multi-dimensional space in a specific way, in which the projection points onto the X_1-X_2, X_2-X_3,…,X_i-X_j axes are equidistant (Fig. 1).

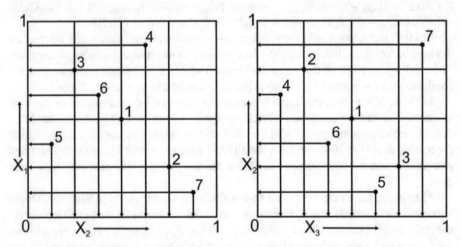

Fig. 1. The experimental test points' projection on the examples of axes' pairs.

The values of the cutting parameters and ones regarding the emulsion mist formation according to the PSI method can be seen in Table 1.

3   Results and Discussion

The values of the individual components of the total force F, in dry turning of AISI 1045 steel are shown in Fig. 2, when cooling by the MQC method in Fig. 3 and, with the use of

Table 1. Cutting parameters' values and parameters of forming the active medium for the individual points in the PSI method

Point according to PSI	Cutting parameters		Parameters of forming the active medium	
	v_c [m/min]	f [mm/rev]	E [g/min]	P [l/min]
1	210	0.167	0.482	2.9
2	105	0.25	0.235	4.9
3	295	0.1	0.706	2.4
4	420	0.2	0.122	1.2
5	150	0.05	0.595	3.9
6	295	0.15	0.348	5.8
7	70	0.3	0.824	2.4

the MQC + EP/AW method – in Fig. 4. The highest values of the Fc cutting force have been obtained for all three methods of cooling. Analysis of the values of the individual components according to the PSI method (Fig. 2, 3 and 4) has shown that the largest influence on all the components is that of feed. For the higher feed per revolution values and low values of cutting speeds (item 2 and item 7 of the PSI method), the highest values have been found, which is a result of the growth of the section of the cut layer and higher tool-workpiece friction.

In an analysis of the application of the selected cooling methods in turning of the AISI 1945 steel, it has been found that the largest percentage differences for the individual components of the total force were those in the cases larger feed per revolution values and lower values of cutting speeds. The less influence of the emulsion mist in the MQC and MQC + EP/AW methods for higher speeds can be explained by the increase of temperature in the cutting zone and more difficult access of single droplets to the cutting zone [26].

The application of the emulsion mist (MQC method) has resulted in 2% to 11% reduction of the cutting force F_c, and the thrust force F_p, as compared to dry machining for six points of the PSI method. An exception is point 5 ($v_c = 150$ m/min; $f = 0.05$ mm/rev, $E = 0.595$ g/min; $P = 3.9$ l/min) where higher values of all the components of the total force have been observed with the application of the MQCL method.

Supply of the emulsion mist with the addition of EP/AW to the cutting zone (Fig. 4) significantly influences reduction of the total force components as compared to the other cooling methods. Only in point 4, increase of the total force components, as compared to the MQC method, has been observed, which can result from the higher cutting speed and low intensity of the air flow. Application of an additive based on phosphate ester results in formation of a lubricating film on the cutting edge [27] which film can separate the friction surfaces and reduce friction in the tool-workpiece interface. During turning in the MQC + EP/AW method, a lubricating film is created in the contact zone and, consequently, the friction coefficient on the rake face decreases and the total force components are reduced. The droplet diameters have a great influence on the total force during turning in the MQC + EP/AW method. For a small flow intensity ($P = 1.2$ l/min – point 4), cooling by emulsion mist with a phosphate ester based additive does not influence the force reduction; on the other hand, when the flow is increased ($P = 4.9$ l/min – point 6), it reduces the total force, F, by 16% and, for maximum flow intensity ($P = 5.8$ l/min – point 6), by as much as 25.4% as compared to dry machining.

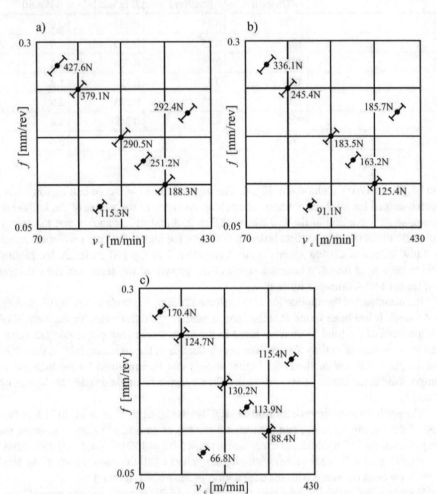

Fig. 2. Values of the F total force components in turning of AISI 1045 steel under the conditions of dry machining for: a) F_c; b) F_f; c) F_p.

The results of tests concerning the total force in turning of AISI 1045 steel show that the selected cooling methods applied in the experiments have significant influence the character of the phenomena influencing the components of the cutting force. The test results have not proved the relationships between the total force components in turning under the MQL conditions and cool machining as presented in literature. It has been stated in the scientific works [20] that the application of machining with minimized

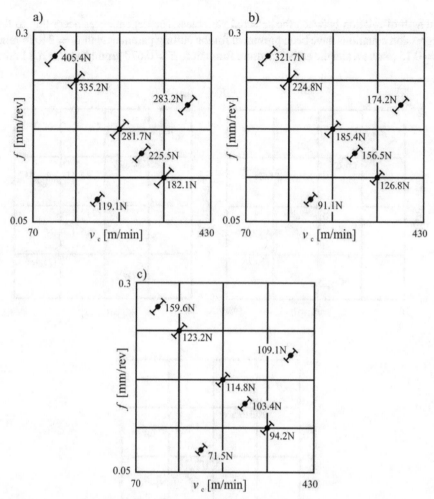

Fig. 3. Values of the F total force components in turning of AISI 1045 steel under the conditions of cooling by the MQC method for: a) F_c; b) F_f; c) F_p.

lubrication does not influence reduction of the total force as compared to dry machining. The authors, however, do not take into consideration the parameters of creating the oil mist in the MQL method and the emulsion mist in the MQC method. The presented results (Fig. 5) allow us to observe how important is the selection of adequate values of the parameters of creating the emulsion mist, especially the air flow intensity [11, 18], which influences the distribution of droplets to the cutting zone and, consequently, changes the conditions of cooling and lubrication in the cutting zone. The values of the cutting power P_e, depending on the cooling technique, based on the results of statistical processing (Table 2) which had been obtained with the use of the Statistica program, have been presented in Fig. 6. In Table 2, one can also find the coefficient of multi-dimensional correlation R, as well as the Cochran index G which inform about the strength of relation between the selected variables. The dependences according to the regression equations have been presented for the cutting parameters of $v_c = 240$ m/min, $f = 0.15$ mm/rev. and the active medium formation, $E = 0.678$ g/min and $P = 5.8$ l/min.

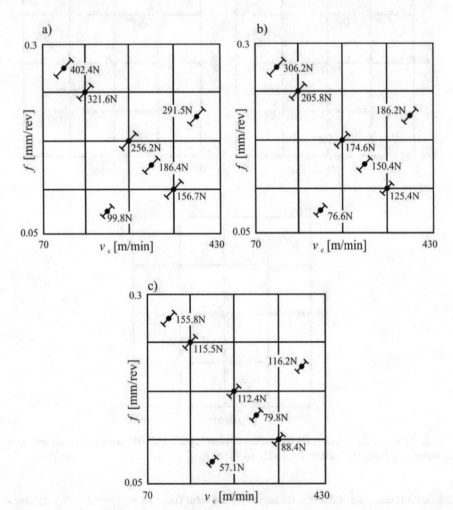

Fig. 4. Values of the F total force components in turning of AISI 1045 steel under the conditions of cooling by the MQC + EP/AW method for: a) F_c; b) F_f; c) F_p.

The test results (Fig. 6) show significant effect of the MQC + EP/AW method on the cutting power in the investigated range. The application of EP/AW technique affects the reduction of cutting power in turning of the AISI 1045 steel by 23% to 35%, comparing to dry machining and by 8% to 23%, in comparison to the MQCL method depending on the variable intensity of the air flow P.

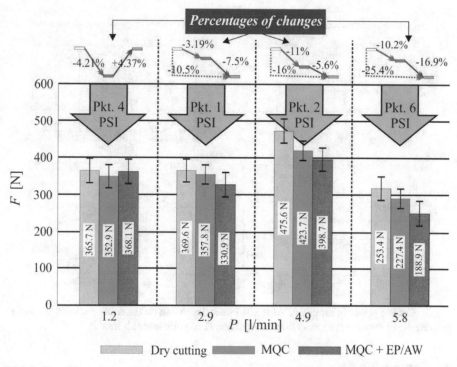

Fig. 5. The effect of the volumetric air flow intensity on the percentage changes of the total force during AISI 1045 steel's turning considering the cooling conditions for selected points of the PSI method.

Table 2. Empirical models for the value of cutting power in turning of AISI 1045 steel

Dry cutting		MQC		MQC + EP/AW	
$P_e = 2,39 \cdot v_c^{0,087} \cdot f^{0,099}$		$P_e = 2,28 \cdot v_c^{0,153} \cdot f^{0,197} \cdot$ $E^{0,054} \cdot P^{-0,124}$		$P_e = 2,13 \cdot v_c^{0,152} \cdot f^{0,195} \cdot$ $E^{0,045} \cdot P^{-0,211}$	
$G = 0,59$	$R = 0,794$	$G = 0,32$	$R = 0,989$	$G = 0,47$	$R = 0,957$

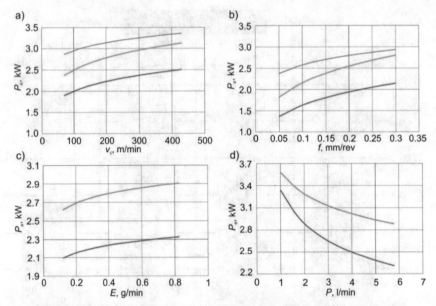

Fig. 6. Cutting power in turning of AISI 1045 steel depending on the cooling conditions for: a) variable v_c; b) variable f; c) mass flow of emulsion E, d) volumetric air flow P.

4 Conclusion

The tests presented in this work have shown differences in the results obtained in turning of AISI 1045 steel depending on the applied cooling method and the conclusions have been formulated as follows:

1. Among the parameters of emulsion mist formation when turning AISI 1045 steel, the cutting force and power are most influenced by the variable intensity of the volumetric air flow P. Smaller droplets obtained for the maximum intensity of the air flow ($P = 5.8$ l/min) better penetrate the cutting zone and, consequently, cause reduction of the cutting force in the zone of the contact of the cutting edge- material under machining-chip.
2. Addition of phosphate ester and calcium to the active medium in the MQC + EP/AW method has resulted in improvement of the lubricating properties of the emulsion by direct influence of chemically active compounds in the process of chip formation and, consequently, significant reduction of the cutting power P_e.
3. The results concerning the cutting power and the total force prove the importance of the selection of adequate values of the parameters, both those the active medium formation and the parameters of cutting.
4. Due to the high number of variables (cutting speed, feed rate, volumetric air flow intensity, mass flow of emulsion) applied in the work to determine the correlation of between the variables, the PSI method has been used, which has allowed for significant reduction of the test points.

Acknowledgments. The financial support from the program of the Polish Minister of Science and Higher Education under the name "Regional Initiative of Excellence" in 2019–2022, project no. 003/RID/2018/19, funding amount 11 936 596.10 PLN.

References

1. Mia, M., Bashir, M.Al., Khan, M.A., Dhar, N.R.: Optimization of MQL flow rate for minimum cutting force and surface roughness in end milling of hardened steel (HRC 40). Int. J. Adv. Manuf. Technol. **89**(1–4), 675–690 (2017)
2. Stachurski, W., Sawicki, J., Wójcik, R., Nadolny, K.: Influence of application of hybrid MQL-CCA method of applying coolant during hob cutter sharpening on cutting blade surface condition. J. Clean. Prod. **171**, 892–910 (2018)
3. Maruda, R.W., Feldshtein, E., Legutko, S., Krolczyk, G.M.: Analysis of contact phenomena and heat exchange in the cutting zone under minimum quantity cooling lubrication conditions. Arab. J. Sci. Eng. **41**(2), 661–668 (2016)
4. Mia, M., Singh, G., Gupta, M.K., Sharma, V.S.: Influence of Ranque-Hilsch vortex tube and nitrogen gas assisted MQL in precision turning of Al 6061-T6. Precis. Eng. **53**, 289–299 (2018)
5. Emami, M., Sadeghi, M.H., Sarhan, A.A.D., Hasani, F.: Investigating the Minimum Quantity Lubrication in grinding of Al_2O_3 engineering ceramic. J. Clean. Prod. **66**, 632–643 (2014)
6. Mia, M., Dhar, N.R.: Optimization of surface roughness and cutting temperature in high-pressure coolant-assisted hard turning using Taguchi method. Int. J. Adv. Manuf. Tech. **88**(1–4), 739–753 (2017)
7. Sieniawski, J., Nadolny, K.: The effect upon grinding fluid demand and workpiece quality when an innovative zonal centrifugal provision method is implemented in the surface grinding of steel CrV12. J. Clean. Prod. **113**, 960–972 (2016)
8. Shokoohi, Y., Khosrojerdi, E., Rassolian Shiadhi, B.H.: Machining and ecological effects of a new developed cutting fluid in combination with different cooling techniques on turning operation. J. Clean. Prod. **94**, 330–339 (2015)
9. Zhang, S., Li, J.F., Wang, Y.W.: Tool life and cutting forces in end milling Inconel 718 under dry and minimum quantity cooling lubrication cutting conditions. J. Clean. Prod. **32**, 81–87 (2012)
10. Maruda, R.W., Feldshtein, E., Legutko, S., Krolczyk, G.M.: Research emulsion mist generation in the conditions of minimum quantity cooling lubrication (MQCL). Teh. Vjesn. **22**(5), 1213–1218 (2015)
11. Park, K., Olortegui-Yume, J., Yoon, M.-C., Kwon, P.: A study on droplets and their distribution for minimum quantity lubrication (MQL). Int. J. Mach. Tool. Manuf. **50**(9), 824–833 (2010)
12. Maruda, R.W., Krolczyk, G.M., Feldshtein, E., Pusavec, F., Szydlowski, M., Legutko, S., Sobczak-Kupiec, A.: A study on droplets sizes, their distribution and heat exchange for minimum quantity cooling lubrication (MQCL). Int. J. Mach. Tool. Manuf. **100**, 81–92 (2016)
13. Hadad, M.J., Tawakoli, T., Sadeghi, M.H., Sadeghi, B.: Temperature and energy partition in minimum quantity lubrication-MQL grinding process. Int. J. Mach. Tool. Manuf. **54–55**, 10–17 (2012)
14. Tawakoli, T., Hadad, M.J., Sadeghi, M.H.: Influence of oil mist parameters on minimum quantity lubrication – MQL grinding process. Int. J. Mach. Tool. Manuf. **50**(6), 521–531 (2010)
15. Iskandar, Y., Tendolkar, A., Attia, M.H., Hendrick, P., Damir, A., Diakodimitris, C.: Flow visualization and characterization for optimized MQL machining of composites. CIRP Ann. - Manuf. Technol. **63**(1), 77–80 (2014)

16. Duchosal, A., Werda, S., Serra, R., Courbon, C., Leroy, R.: Experimental method to analyze the oil mist impingement over an insert used in MQL milling process. Measurment **86**, 284–292 (2016)
17. Obikawa, T., Asano, Y., Kamata, Y.: Computer fluid dynamics analysis for efficient spraying of oil mist in finish-turning of Inconel 718. Int. J. Mach. Tool. Manuf. **49**(12–13), 971–978 (2009)
18. Cabanettes, F., Faverjon, P., Sova, A., Dumont, F., Rech, J.: MQL machining: from mist generation to tribological behavior of different oils. Int. J. Adv. Manuf. Techn. **90**(1–4), 1119–1130 (2017)
19. Królczyk, G.M.: Morfologia powierzchni stali duplex po procesie toczenia na sucho i z chłodzeniem. Wydawnictwo Politechniki Opolskiej (2015)
20. Kaynak, Y., Karaca, H.E., Noebe, R.D., Jawahir, I.S.: Tool-wear analysis in cryogenic machining of NiTi shape memory alloys: a comparison of tool-wear performance with dry and MQL machining. Wear **306**(1–2), 51–63 (2013)
21. Kaynak, Y., Karaca, H.E., Noebe, R.D.., Jawahir, I.S.: The effect of active phase of the work material on machining performance of a NiTi shape memory alloy. Metall Mater Trans A, **46**(6), 2625–2636 (2015)
22. Behera, B.C., Ghosh, S., Rao, P.V.: Modeling of cutting force in MQL machining environment considering chip tool contact friction. Tribol. Int. **117**, 283–295 (2018)
23. Pervaiz, S., Deiab, I., Rashid, A., Nicolescu, M.: Minimal quantity cooling lubrication in turning of Ti6Al4V: influence on surface roughness, cutting force and tool wear. Proc. Inst. Mech. Eng. Part B J. Eng. Manuf. **18**(1), 5–14 (2017)
24. Statnikov, R.B., Statnikov, A.: The Parameter Space Investigation Method Toolkit. Artech House, Boston/London (2011)
25. Leksycki, K., Feldshtein, E.: The geometric surface structure of X5CrNiCuNb16-4 stainless steel in wet and dry finish turning conditions. In: Diering, M., Wieczorowski, M., Brown, C.A. (eds.) Manufacturing: International Scientific-Technical Conference Manufacturing: Advance Manufacturing II, vol. 5, pp. 183–194 (2019)
26. Diniz, A.E., De Oliveira, A.J.: Optimizing the use of dry cutting in rough turning steel operations. Int. J. Mach. Tool. Manuf. **44**(10), 1061–1067 (2004)
27. Maruda, R.W., Krolczyk, G.M., Wojciechowski, S., Zak, K., Habrat, W., Nieslony, P.: Effects of extreme pressure and anti-wear additives on surface topography and tool wear during MQCL turning of AISI 1045 steel. J. Mech. Sci. Techn. **32**(4), 1585–1591 (2018)

Surface Roughness of SMA NiTi in Turning Tests

Anna Zawada-Tomkiewicz[(⊠)], Radosław Patyk, Piotr Socha, and Mateusz Polilejko

Koszalin University of Technology, Koszalin, Poland
anna.zawada-tomkiewicz@tu.koszalin.pl

Abstract. The aim of the paper was to identify the physical phenomena in machining processes of shape memory alloys (SMA) NiTi. The possibilities of machining NiTi alloys are limited due to the low and variable elastic modulus for phases and other material properties that place it in the range of hard-to-machine materials. The article presents the impact of cutting speed on surface roughness. The analysis was carried out in the range of low cutting speeds with the permissible value of tool wear. The research was carried out using averaged power spectrum density, parametric evaluation and the Stockwell transform. Averaged power spectrum density surface profile revealed the appearance of low-frequency components in the wavelength range from 0.3 mm to 1 mm, the amplitude of which increased with increasing cutting speed. A parametric surface roughness analysis was carried out, where a set of surface parameters was determined for 25 measured profiles. In this way, changes during cutting could be observed. Nine surface parameters were determined every 0.6 mm over the length of the shaft and every 15° on the revolution. For each parameter, the intensity of changes was determined, which, collected for all parameters, were compiled in the form of a bar chart. At the same time, the profile was analyzed both in time and frequency. The result of the Stockwell transformation shows how the surface roughness signal changed simultaneously both in time (equivalent to parametric analysis) and frequency (equivalent to frequency analysis in the frequency range 1000 1/m to 3333 1/m, which corresponds to a wavelength from 0.3 mm to 1 mm).

Keywords: Turning · Shape memory alloys · Surface roughness · Stockwell transform

1 Introduction

The main research issue of the paper is the study of physical phenomena in machining processes of shape memory alloys (SMA) NiTi alloys in the context of machined surface creation. The shape memory phenomenon is based on the first kind of martensitic transformation, which occurs through the nucleation of a new phase and its growth. After transformation, martensite has the same chemical composition, degree of atomic ordering and defect of the crystal lattice as the parent phase.

The possibilities of machining of NiTi alloys are limited due to the low and variable modulus of elasticity for phases and other material properties that place it in the range

© Springer Nature Switzerland AG 2020
G. M. Królczyk et al. (Eds.): IMM 2019, LNME, pp. 163–171, 2020.
https://doi.org/10.1007/978-3-030-49910-5_15

of hard-to-machine materials. Understanding the mechanisms and understanding the physical phenomena in the cutting of these alloys will allow the creation of new models, both physical and mathematical, that enable the wider use of shape memory alloys in machine construction.

The current state of knowledge in the field of machining of shape memory alloys indicates a great interest in this type of materials with very modest studies in the field of their machining. The NiTi alloys have been known since the sixties of the last century and in many centers around the world the research of thermo-mechanical properties of shape memory alloys have been conducted [1, 2]. NiTi alloys possess such features as biocompatibility, high ductility, high strength to weight ratio, good fatigue and corrosion resistance, and high damping capacities [3]. They are excellent in exploitation but they seem to be difficult to process due to their high ductility, temperature sensitivity and strong local material hardening [6]. When machining of NiTi alloys the large amounts of the material is not separated from the work-piece and appear as burrs [4, 5]. In the last few years, attempts have been made to process NiTi alloys with conventional machining. These were studies carried out as part of NASA EPSCOR Program and NASA FAP Aeronautical Sciences Project [6]. One of the main problems identified in these studies on the machining of NiTi alloys was, apart from the rapid wear of the tool, also the quality of the machined surface. Observed physical phenomena of difficult-to-cut materials in machining was the result of their physical properties such as super-elasticity and phase change during machining. Machining in finishing operations is even more challenging due to the size effect, which drives material springback for ductile phases, and ploughing effects when undeformed chip thickness is comparable to the cutting edge radius or lower [7].

The paper presents the thorough analysis of surface roughness for a variable cutting speed in the dry cutting process of SMA NiTi alloys. The machined surface was measured and analyzed in 2D arrangement due to the fact that the 2D of the surface is the most representative for a turned surface. The research embraces the surface roughness parametric analysis in time, the averaged power spectral density analysis in frequency and the Stockwell transform for time-frequency analysis. The analysis performed in this way enable the full analysis of a machined surface profile. Analysis of this kind is necessary due to the fact that there are many effects when cutting the SMA NiTi alloy, which adversely affect surface roughness.

2 Materials and Methods

The machining process was carried out under dry conditions. The cutting conditions are presented in Table 1. Turning tests were performed on a high rigidity NEF 400 lathe (Fig. 1). The annealed NiTi alloy rod (ASTM F2063) 90 mm long with an external diameter of 18 mm was used. The aim of this study was the assessment of the surface roughness achieved during the process of turning with coated sintered carbide tools. The machined surface was measured in the 2D arrangement using contact technique. A set of the 2D roughness parameters was determined by performing roughness measurements. For roughness parameters the cut-off was set as 0.8 mm with Gaussian filter.

Table 1. Cutting conditions.

Cutting speed, m/min	Feed rate, mm/rev.	Depth of cut, mm	Cutting distance, mm
12.5; 25.0; 37.5; 50.0; 62.5; 75.0	0.10	0.3	15

Fig. 1. Cutting of the SMA NiTi.

3   Machined Surface Analysis

The surface roughness of machine parts is assessed in finishing operations, because it is decisive for such properties of the surface as the visual aspect of the texture, friction and wear. Generally, the smaller the roughness value, the smoother and more favorable is the surface. Parametric analysis of the surface is used most often and refers, for example, to such aspects as

- selection of signal pre-processing methods, filtration methods, cut-off [8],
- selection of surface parameters from the set of possible and their evaluation due to their classification bias, dispersion of values, measurement possibilities and others [9, 10],
- evaluation of surfaces difficult to measure: structural or free-form surfaces [11].

The surface roughness analyses for turning operation focus on the kinematic-geometrical mapping of the tool nose. The radius of the tool nose determines an almost flat stretch parallel to the surface of a length similar to the value of the feed rate, particularly when the feed rate is very small. The theoretical maximal roughness, calculated for cutting conditions, is estimated to be 3.1 μm. The measured roughness has been higher than the theoretical value (respectively for 12.5 m/min – 1.52×; 25.0 m/min – 1.12×; 37.5 m/min – 1.51×; 50.0 m/min – 2.24×; 62.5 m/min – 2.25×; 75.0 m/min – 19.29×). There are several reasons for this phenomenon [12].

- The first one is the consequence of poor thermal conductivity of NiTi alloy, which does not allow sufficient removal of heat from the machining zone. It was manifested by continuous accelerated wear of the tool and blunting of the cutting edge. Tool wear affects the minimal irremovable layer of the cut material (undeformed chip thickness).

It follows that an increase in the radius of the edge of the blade and blunting of the cutting edge causes an increased effect of side flow of the material, and thus an increase in the amplitude of surface height low frequencies. In addition, the tool wear is associated with cutting edge chipping, whose envelope reproduces in the range of high spatial frequencies of the surface. When cutting with technical edges featuring a non-zero edge radius in the conditions of constrained cutting, the side flow of the machined material causes increased roughness.

- In the process of SMA NiTi cutting a several other features influence the surface roughness. These include shape memory effect and superelasticity. These features combined with the low thermal conductivity and ductility of the material affect chip breaking process and the formation of burrs. All these factors negatively affect the surface texture, which is distant from the model resulting from the kinematics of the process.

The most favorable cutting speed for which surface roughness Ra parameter has been less than 1 μm should be sought among the first three points of the experiment, where the maximum roughness ratio, in relation to the theoretical, is less than 2. The least favorable result of the average surface roughness Ra was obtained for the value of cutting speed 75 m/min. The value of Ra corresponds to the surface roughness range as for roughing.

To explain the increase in parameter values with the increase in cutting speed the frequency components included in the roughness profile wear analysed, Fig. 2 summarizes the averaged power spectrum density as a function of wavelength. For a cutting speed of 12.5 m/min, the surface profile is dominated by feed rate (region B). This is evident in the form of the dominant maximum for a wavelength of 0.1 mm. The wear of the blade for the machined material volume of 28 mm^3 did not exceed 15% of the limit value. The finest component of surface texture composed of high frequencies is the microroughness (region B). This component is very low in energy and its influence on roughness parameters is small. Vibrations appeared in the process for higher wavelengths (region C), but their value was small for cutting speed 12.5 m/min and did not dominate the profile height.

For a cutting speed of 25 m/min (Fig. 2b), with a blade wear of 30% of the allowable value, the kinematic and geometrical mapping of the shape of the blade corner in the workpiece is dominated by vibrations, which in this case have smaller amplitude than for a speed of 12.5 m/min. The distribution of profile heights is the most favorable when considering such roughness parameters as: Ra, Rt, Rp, Rv, Rsk, Rku and others but the surface profile is irregular.

Increasing the cutting speed to 37.5 m/min not only caused an increase in tool wear for the same volume of material removed, but as shown in Fig. 2c, caused the appearance of vibration in the system and ultimately increased the distribution of profile height and increased value of most surface parameters.

Figure 2 does not show whether and how many changes occurred in the signal. The averaged power spectrum density of roughness does not indicate how the surface changes during cutting. The Fourier transform is a useful tool in the study only of stationary signals. In order to record changes during cutting, surface roughness parameters were measured and determined independently in accordance with the measurement experiment. The shaft surface was measured in 25 measuring lengths. Each subsequent measurement was carried out with a shift of 0.6 mm and shaft rotation by 15°. Finally,

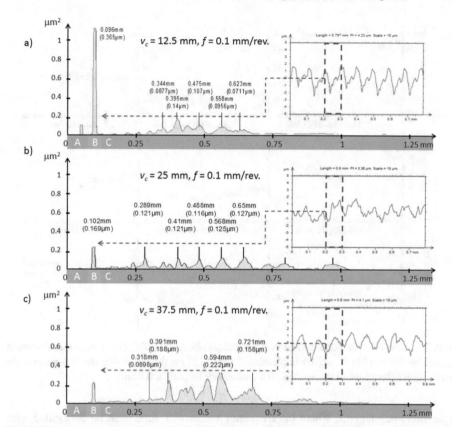

Fig. 2. Averaged power spectrum density of surface roughness for a measuring length of 15 mm for cutting speed a) 12.5 m/min, b) 25 m/min, c) 32.5 m/min with marking the feed on the elementary length of the profile and indicating the appropriate wavelength corresponding to the feed.

25 measured profiles with a length of 4 mm were analyzed and a set of parameters was obtained, of which three examples of *Ra, Rt* and *Rsk* for three cutting speeds are presented in Fig. 3. For a cutting speed of 12.5 m/min it can be distinguished in the range of these three parameters from 4 to 6 changes. For higher cutting speeds the surface profile is more diverse and the changes are much more visible.

The total intensity of changes in 9 parameters *Ra, Rt, Rp, Rv, Rsk, Rku, Rfd, Af, CH* in time was shown in Fig. 4 as a bar chart plotted on the roughness profile. The scale of the bar chart $1 \div 9$ corresponds to the sum of positively classified measurement points at which there is a probability that the next one differs statistically from the current one.

Observation and analysis of the profile alone do not allow changes to be observed. Parametric analysis has this limitation that the parameters are determined for the measuring length. Although changes in the content of spatial frequencies allow observation of changes in the profile as a whole, they do not indicate the time of their appearance, because the Fourier transform does not contain temporal information. The analysis of

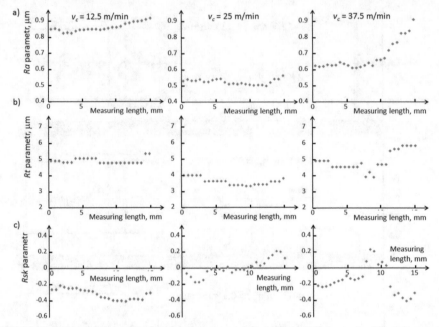

Fig. 3. The value of the machined surface parameter a) *Ra*, b) *Rt*, c) *Rsk*, determined for subsequent points on the machined surface implemented every 15° on the revolution and 0.6 mm on the translation determined as the average of 5 elementary lengths of 0.8 mm.

non-stationary signals, where the frequency varies with time must be performed with the use of appropriate methods [13]. The article uses the Stockwell transform.

4 Discussion on Changes in the Machined Surface

The Stockwell transform allows the analysis of a roughness profile in a length-spatial frequency distribution. The Stockwell transform provides a frequency-dependent resolution and it also maintains a direct relationship with the Fourier transform and its spectrum. Therefore, it obtains local phase information with the referenced time/length equal to zero [14].

Figure 4 shows the spectrum of the Stockwell transform for a measuring length of 15 mm and a spatial frequency of up to 3333 1/m for subsequent cutting speeds. The y-axis represents the spatial frequency distribution and the x-axis represents the measuring length in mm.

The spectrum of the Stockwell transform is presented together with the surface profile and bar chart of changes in surface parameters. As can be seen in the spectrum contour chart, information about changes in the surface roughness profile in this case is much more precise, because these changes can not only be classified in terms of intensity, but refer them more precisely to the measurement time/length and spatial frequency.

For cutting speed 12.5 m/min (Fig. 4a), the intensity of surface profile changes does not exceed 0.2 μm, and the changes are evenly distributed along the measurement length.

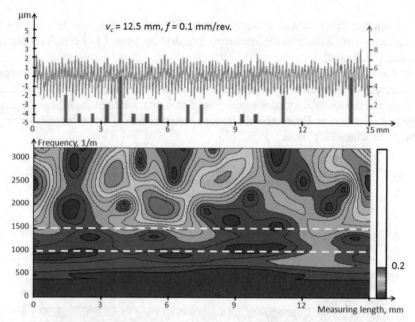

Fig. 4a. Surface roughness profile for a measuring length of 15 mm including the marking of intensity bars for changes in surface parameters and the Stockwell Transform spectrograms for a cutting speed of 12.5 m/min.

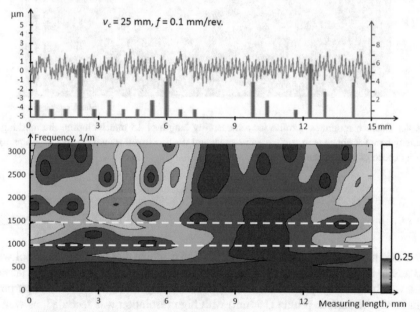

Fig. 4b. Surface roughness profile for a measuring length of 15 mm including the marking of intensity bars for changes in surface parameters and the Stockwell transform spectrograms for a cutting speed of 25 m/min.

The situation is different with a cutting speed of 25 m/min (Fig. 4b). In this case, the changes are visible mainly in the measuring length range from 3 to 7 mm. The intensity of changes is similar.

In the case of higher cutting speed 37.5 m/min (Fig. 4c), changes of high intensity are observed for the section from 13 to 14 mm. However, changes with lower intensity are visible in the lower frequency range from 3 to 4 mm, while in the higher frequency range from 6 to 8 mm. Characteristic for all spectral results is that it indicates changes in the place where they occur.

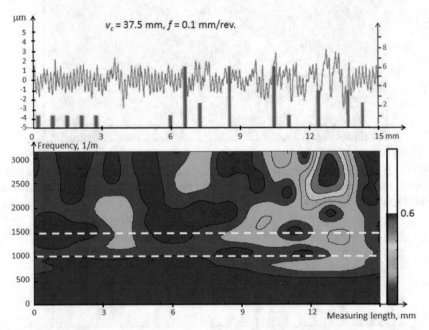

Fig. 4c. Surface roughness profile for a measuring length of 15 mm including the marking of intensity bars for changes in surface parameters and the Stockwell transform spectrograms for a cutting speed of 37.5 m/min.

5   Conclusions

The NiTi shape memory materials are difficult to machine due to accelerated tool wear and unsatisfactory surface quality. The article attempts to analyze the machined surface for a variable cutting speed while testing surface changes over time. For this purpose, both the entire surface profile (15 mm) with high resolution was recorded, as well as subsequent surface profiles for parametric evaluation.

The tests were carried out for six cutting speeds in the range from 12.5 m/min to 75 m/min. This range for the predicted volume of machined material did not cause exceeding the permissible value of blade wear. A constant feed value and cutting depth were adopted. In the examined range, parametric analysis of the surface showed significant differences. The *Ra* parameter alone for three lower cutting speeds was less than 1 μm, for higher cutting speeds it was much higher.

Analysis of surface parameters during cutting showed significant differences, which was confirmed using the Stockwell transform. Thanks to the Stockwell transform, it was possible to quantify changes over the entire frequency and time range. The contour chart of the transform spectrum showed changes in the spatial frequency components over the measurement length. Spectrum analysis shows that for a cutting speed of 12.5 m/min, the surface undergoes constant changes in various frequency ranges. The intensity of these changes is not high. At a cutting speed of 25 m/min, the intensity is still low, but the changes are more orderly. For a cutting speed of 37.5 m/min, the change of intensity increases, which is 2 to 3 times higher than that for lower cutting speeds.

References

1. Huang, X., Liu, Y.: Effect of annealing on the transformation behavior and superelasticity of NiTi shape memory alloy. Scr. Mater. **45**(2), 153–160 (2001)
2. Saadat, S., Salichs, J., Noori, M., Hou, Z., Davoodi, H., Bar-on, I., Suzuki, Y., Masuda, A.: An overview of vibration and seismic applications of NiTi shape memory alloy. Smart Mater. Struct. **11**, 218–229 (2002)
3. Kong, M.C., Axinte, D., Voice, W.: Challenges in using waterjet machining of nickel titanium alloy shape memory alloys: an analysis of controlled-depth milling. J. Mater. Process. Technol. **211**(6), 959–971 (2011)
4. Piquard, R., D'Acunto, L.P., Dudzinski, D.: Micro-end milling of nickel titanium alloy biomedical alloys, burr formation and phase transformation. Precis. Eng. **38**(2), 356–364 (2014)
5. Weinert, K., Petzoldt, V.: Machining NiTi micro-parts by micro-milling. Mater. Sci. Eng., A **481**(1), 672–675 (2008)
6. Kaynak, Y., Manchiraju, S., Jawahir, I.S.: Modeling and simulation of machining-induced surface integrity characteristics of NiTi alloy. Procedia CIRP **31**, 557–562 (2015)
7. Kacalak, W., Lipiński, D., Bałasz, B., Rypina, Ł., Tandecka, K., Szafraniec, F.: Performance evaluation of the grinding wheel with aggregates of grains in grinding of Ti-6Al-4 V titanium alloy. Int. J. Adv. Manuf. Technol. **94**, 301–314 (2018)
8. Zawada-Tomkiewicz, A.: Free-form surface data registration and fusion. The case of roughness measurements of a convex surface. Metrol. Meas. Syst. **2**, 323–333 (2019)
9. Lipiński, D., Kacalak, W.: Metrological aspects of abrasive tool active surface topography evaluation. Metrol. Meas. Syst. **23**(4), 567–577 (2016)
10. Zawada-Tomkiewicz, A., Ściegienka, R.: Monitoring of a micro-smoothing process with the use of machined surface images. Metrol. Meas. Syst. **XVIII**(3), 419–428 (2011)
11. Hreha, P., Radvanska, A., Knapčíková, L., Królczyk, G., Legutko, S., Królczyk, J.B., Hloch, S., Monka, P.: Roughness parameters calculation by means of on-line vibration monitoring emerging from AWJ interaction with material. Metrol. Meas. Syst. **XXII**(2), 315–326 (2015)
12. Storch, B., Zawada-Tomkiewicz, A.: Distribution of unit forces on the tool nose rounding in the case of constrained turning. Int. J. Mach. Tools Manuf **57**, 1–9 (2012)
13. Zawada-Tomkiewicz, A., Ściegienka, R., Zuperl, U., Stępień, K.: Images of the machined surface in evaluation of the efficiency of a micro-smoothing process. Strojniški vestnik – J. Mech. Eng. **65**, 7–8 (2019)
14. Lima, É.M., Brito, N.S.D., de Souza, B.A.: High impedance fault detection based on Stockwell transform and third harmonic current phase angle. Electr. Power Syst. Res. **175**, 105931 (2019)

The Investigations of the Surface Layer Properties of C45 Steel After Plasma Cutting and Centrifugal Shot Peening

Kazimierz Zaleski, Agnieszka Skoczylas[✉], and Krzysztof Ciecielag

Faculty of Mechanical Engineering, Department of Production Engineering, Lublin University of Technology, 36 Nadbystrzycka, 20-618 Lublin, Poland
a.skoczylas@pollub.pl

Abstract. The article presents the research of the results 3D roughness parameters and micro-hardness of the surface layer of plasma-cut C45 steel parts after centrifugal shot peening. Centrifugal shot peening research were performed on an FV-580a vertical machining center using a specially designed centrifugal shot peening head. Variable parameters of the centrifugal shot peening were: peripheral speed v_g, feed rate v_f, transverse feed f_p and number of passes i. The use of centrifugal shot peening for finish machining of plasma-cut C45 steel parts allowed maximum reduction of roughness 3D parameters of up 5.3 times. After centrifugal shot peening the microhardness maximum increase by $\Delta HV = 132$ and the thickness of the hardened layer is from 40 μm to 85 μm.

Keywords: Plasma cutting · Centrifugal shot peening · 3D roughness parameters · Microhardness

1 Introduction

The properties of the surface layer of machine elements mostly depend on the method and parameters of finishing machining [1–3]. The condition of the surface layer formed in the production process has an impact on the functional properties of machine elements, such as fatigue strength [4, 5], wear resistance [6, 7], corrosion resistance [8, 9], energy state and adhesive properties [10, 11]. Machining methods that significantly improve the surface layer properties of machine components include burnishing and shot peening. Burnishing, which is characterized by a constant value of the tool interaction force on the workpiece, allows obtaining a small surface roughness [12, 13]. The quality of the obtained surface depends not only on the roughness of the surface before burnishing and the technological parameters used, but also on the machining strategy [14]. Both burnishing and shot peening increases the microhardness of the surface layer of workpieces and create compressive residual stresses in this layer. These changes in the physical properties of the surface layer are associated with an increase in density and a change in the concentration of defects in the crystal structure, which is confirmed by studies conducted with annihilation techniques [15, 16].

© Springer Nature Switzerland AG 2020
G. M. Królczyk et al. (Eds.): IMM 2019, LNME, pp. 172–185, 2020.
https://doi.org/10.1007/978-3-030-49910-5_16

Shot peening, which involves hitting the balls into the machining surface, is characterized by the occurrence of much smaller forces acting on the workpiece compared to burnishing. Therefore, the shot peening may be used as finishing machining for elements with low stiffness. Most often, the shot peening is used to increase the strength and fatigue life of components exposed during operation to variable loads [17, 18]. During shot peening, the balls hitting the machined surface cause "knocking out" of cavities, which the shape and dimensions depend on the workpiece material, the diameter of the balls and technological parameters. During sliding friction, these depressions constitute "tanks" in which the lubricating medium is kept, which reduces wear [19]. Shot peening also has a positive effect on the service life of machine components exposed to material fatigue and tribological wear at the same time [20]. The impact of shot peening on corrosion resistance is related to the type of machining medium and the processing technological parameters [21]. The beneficial effect of shot peening, in addition to improving the properties of the surface layer, can be deburring and rounding the edges of the workpiece [22].

One of the varieties of shot peening is a centrifugal shot peening. It involves hitting the surface with balls that have the ability to move in the radial holes of the rotating head (they are secured against falling out) [23, 24]. Centrifugal shot peening can be integrated with grinding [25].

In previous research, centrifugal shot peening was applied to surfaces machined by milling and grinding. Plasma cutting is one of the commonly used methods for dividing engineering materials. This is a thermal cutting process, and therefore there is a very high temperature in the cutting zone. The cut elements heat up to a certain depth (heat affected zone, which causes changes in the properties of the surface layer of these elements, and their surfaces are contaminated with slag arising during cutting). The surface obtained after plasma cutting is generally characterized by a higher roughness compared to grinding and milling [26].

The aim of this study was to assess the impact of centrifugal shot peening parameters on surface roughness and microhardness of the surface layer of plasma cut steel elements.

2 Research Methodology

In research used samples made of C45 steel (determination in accordance with EN ISO 683-2: 2018). In the machine-building industry, C45 non-alloy steel is used for medium loaded machine and equipment components, e.g. shafts, spindles, axles, unhardened gears. Table 1 presents the chemical composition and selected mechanical properties of the C45 non-alloy steel tested (according to the material card).

Cuboid samples of 5 mm × 8 mm × 100 mm were cut with plasma. For plasma cutting, a plasma cutter powered by Kjellberg HiFocus 80i generator with a working range of 10 ÷ 80 A and equipped with a 3000 × 1500 mm extraction table by Eckert was used. The following process parameters were used for cutting: cutting speed v = 2500 mm/min, gas pressure: 0.8 MPa, current intensity: I = 80 A. Nitrogen was used as plasma gas.

As a finishing machining of the surface after plasma cutting, centrifugal shot peening was used, which was carried out on the vertical machining center FV - 580a. A special

Table 1. Chemical composition and selected properties of C45 steel

Chemical composition, [%]								
C	Mn	Si	P	S	Cr	Ni	Mo	Fe
0.48	0.74	0.36	0.011	0.01	0.09	0.02	0.002	Other
Yield strength				$R_e = 430$ MPa				
Tensile strength				$R_m = 740$ MPa				
Hardness				250 HB				

tool was used for shot peening, which was a centrifugal shot peening head with an outer diameter of $D_g = 70$ mm. The head has symmetrically arranged balls with a diameter of $d_k = 6.3$ mm in the number $z_k = 12$ (Fig. 1), which after starting the rotary movement of the tool, under the influence of centrifugal force hit the surface of the workpiece. During centrifugal shot peening, the sample (2) was clamped in the chuck, which was in the jaws of the vice mounted on the machine table, moving at a speed of v_f. The centrifugal shot peening head (1) performed rotational motion at the speed n and simultaneously the transverse feed f_p. After starting the rotation of the head at speed n, the balls hit the surface of the sample (2). During impact, the ball bounces off the work surface and is retracted deep into the seat at a distance of g, called the head infeed to the workpiece.

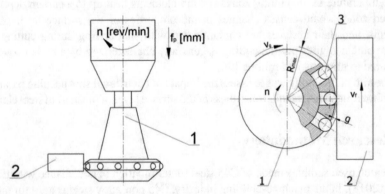

Fig. 1. Scheme of the centrifugal shot peening on a vertical machining center (1 – centrifugal shot peening head, 2 – sample, 3 – ball (g – infeed, R_{max} – maximum head radius)

Centrifugal shot peening was carried out using the Mobile Cut cooling lubricant, with a constant infeed value of $g = 0.5$ mm. Variable parameters of the centrifugal shot peening process were:

– peripheral speed: $v_g = 681 \div 939$ m/min
– feed rate: $v_f = 1368 \div 5928$ mm/min
– transverse feed: $f_p = 0.03 \div 0.13$ mm
– number of passes: $i = 1 \div 3$.

3D topography measurements were made using the Hommel - Etamic T8000RC 120-140 device. The 3 mm × 3 mm area was scanned. The same area was adopted for the surface after plasma cutting as after plasma cutting and centrifugal shot peening. Using the HommelMap Basic software, 3D surface roughness parameters were determined.

Microhardness measurements were made using the Vickers method in accordance with EN ISO 6507-1: 2018 on sections after standard preparatory machining. The LM 700 AT microhardness tester was used, using an indenter load of 50 g (HV 0.05). Based on the results obtained from the microhardness measurements, the increase in hardness ΔHV and the thickness of the layer hardened with centrifugal shot peening g_h was determined (Fig. 2).

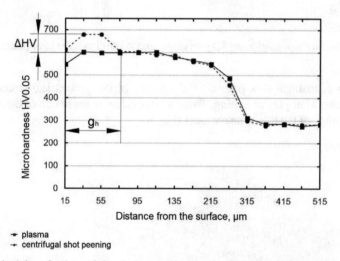

Fig. 2. Methodology for determining the increase in microhardness and thickness of the reinforced layer

3 Research Results

Figure 3 shows the surface topography after plasma cutting. The surface after thermal cutting is characterized by the presence of numerous elevations and depressions. On the surface there are numerous craters and a ribbed structure, the occurrence of which may indicate the oscillation of the plasma head and the fluctuation of the plasma beam during the cutting process. The presence of numerous depressions and elevations on the cut surface can be explained by the fact that the linear burning rate of the material is greater than the velocity of the plasma beam.

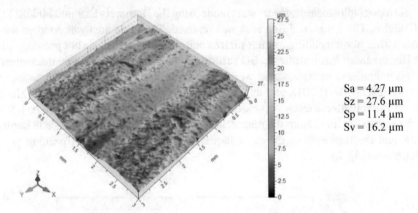

$Sa = 4.27 \ \mu m$
$Sz = 27.6 \ \mu m$
$Sp = 11.4 \ \mu m$
$Sv = 16.2 \ \mu m$

Fig. 3. Surface topography and 3D parameters for C45 steel samples after plasma cutting

During centrifugal shot peening, deformation of the geometrical structure of the surface occurs after plasma cutting. There is a reduction in the height of the fringes and the "flattening" of the micro-unevenness (Fig. 4).

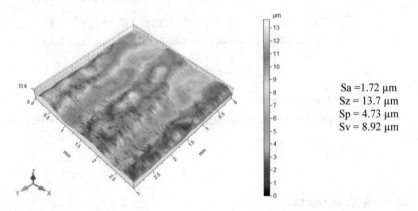

$Sa = 1.72 \ \mu m$
$Sz = 13.7 \ \mu m$
$Sp = 4.73 \ \mu m$
$Sv = 8.92 \ \mu m$

Fig. 4. Surface topography and 3D parameters for C45 steel samples after plasma cutting and centrifugal shot peening ($v_g = 989$ m/min, $v_f = 3648$ mm/min, $f_p = 0.08$ mm, i = 1)

During centrifugal shot peening, an increase in peripheral speed in the range of $v_g = 681 \div 989$ m/min causes a decrease in surface roughness (Fig. 5). This is due to more intense deformation and reduction of the fringes height due to the increasing impact energy. After exceeding the speed $v_g = 835$ m/min, increase in the roughness parameters Sa and Sz can be seen. The use of circumferential speed, greater than $v_g = 835$ m/min, causes intensive plastic deformation of the processed surface. Numerous depressions and elevations arise that cause deterioration of the surface quality. The type of the changes for the roughness parameter Sp (Fig. 6) as a function of peripheral speed v_g is the same as for the parameters Sa and Sz. On the other hand, the increase in peripheral velocity causes the occurrence of more depressions on the shot peening surface, which confirms the increase in the parameter Sv. These depressions may be potential lubricant reservoirs. The roughness parameters Sa, Sz, Sp and Sv decreased from 1.7 to 4.2 times relative to the surface after plasma cutting.

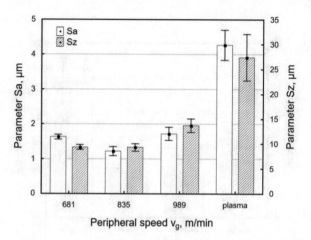

Fig. 5. Influence of peripheral speed v_g on Sa and Sz roughness parameters of plasma cut and centrifugally shot peening specimens ($v_f = 3648$ mm/min, $f_p = 0.08$ mm, $i = 1$)

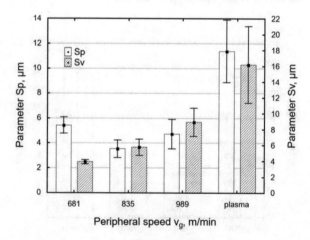

Fig. 6. Influence of peripheral speed v_g on Sp and Sv roughness parameters of plasma cut and centrifugally shot peening specimens ($v_f = 3648$ mm/min, $f_p = 0.08$ mm, $i = 1$)

Figure 7 presents the impact of feed rate v_f on the roughness parameters Sa and Sz. The increase in the feed rate results in a decrease in the number of ball hits per unit of surface, which results in a slight increase in the roughness parameter Sa. In the case of the Sz parameter for v_f in the range of 1368 mm/min to 3648 mm/min the analyzed parameter is reduced. Parameters Sa and Sz decreased from 1.8 to 3.8 times relative to the value after plasma cutting. For the parameter Sp (maximum height of surface elevations), the type of the changes is the same as for the parameter Sz. The maximum values for depressions of the surface decrease as the speed of workpiece movement increases (Fig. 8).

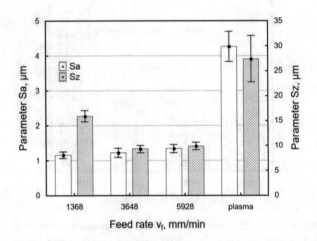

Fig. 7. Influence of feed rate v_f on Sa and Sz roughness parameters of plasma cut and centrifugally shot peening specimens $(v_g = 835$ m/min, $f_p = 0.08$ mm, $i = 1)$

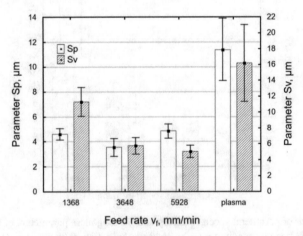

Fig. 8. Influence of feed rate v_f on Sp and Sv roughness parameters of plasma cut and centrifugally shot peening specimens $(v_g = 835$ m/min, $f_p = 0.08$ mm, $i = 1)$

Increasing the transverse feed f_p, and thus reducing the density of centrifugal shot peening, causes an increase in surface roughness parameters Sa, Sz, Sp (Fig. 9 and Fig. 10). Tracks of ball impacts appear at a greater distance from each other, causing uneven deformation of the machined surface. In the range of transverse feed $f_p = 0.08 \div$ 0.13 mm, an intensive increase in the maximum height of surface elevations (parameter Sp) is visible, which is probably caused by the "incomplete" deformation of the micro-unevenness after plasma cutting. The increase in transverse feed f_p causes depressions with a smaller depth on the machining surface, which may limit the lubricant's retention capacity by the surface (Fig. 10).

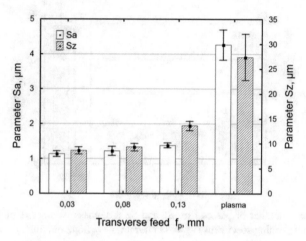

Fig. 9. Influence of transverse feed f_p on Sa and Sz roughness parameters of plasma cut and centrifugally shot peening specimens ($v_g = 835$ m/min, $v_f = 3648$ mm/min, $i = 1$)

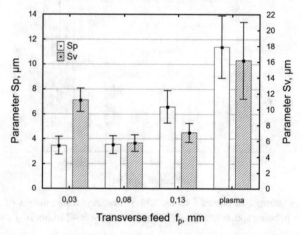

Fig. 10. Influence of transverse feed f_p on Sp and Sv roughness parameters of plasma cut and centrifugally shot peening specimens ($v_g = 835$ m/min, $v_f = 3648$ mm/min, $i = 1$)

Surface roughness is also affected by the number of centrifugal shot peening passes (Fig. 11 and Fig. 12). The increase in the number of passes i causes the machining tracks to partially overlap, which contributes to repeated deformation of the same micro-unevenness. The result is a surface with less roughness. The surface roughness parameters Sa and Sz decreased from 2.9 times to 5.3 times relative to the value after plasma cutting, while the dynamics of changes between the second and third pass decreases.

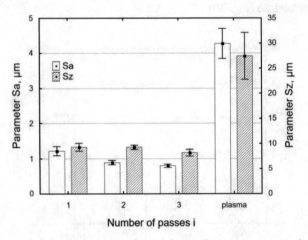

Fig. 11. Influence number of passes i on Sa and Sz roughness parameters of plasma cut and centrifugally shot peening specimens ($v_g = 835$ m/min, $v_f = 3648$ mm/min, $f_p = 0.08$ mm)

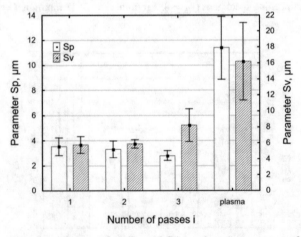

Fig. 12. Influence number of passes i on Sp and Sv roughness parameters of plasma cut and centrifugally shot peening specimens ($v_g = 835$ m/min, $v_f = 3648$ mm/min, $f_p = 0.08$ mm)

The increase in the number of centrifugal shot peening passes i causes that the peaks of micro-unevenness after plasma cutting are more "flattened", which translates into a decrease in the value of the roughness parameter Sp. At the same time, "punching" occurs on the centrifugally shot peening surface of deep depressions as a result of repeated contact of the balls with the workpiece, which translates into an increase in the parameter

Sv (Fig. 12). Changes in the roughness parameters Sp and Sv are more noticeable between the second and third centrifugal shot peening pass.

After plasma cutting, the width of the zone in which the micro-hardness increases is over 350 μm. The microhardness of the surface layer increases in the area of the cutting edge more than twice and amounts to approx. 510 HV.

Changes in the microhardness of the surface layer of C45 steel after plasma cutting and centrifugal shot peening depend on the technological parameters of the process. An increase in the peripheral speed of the head v_g causes an increase in the number of impacts per unit area, which results in a greater proportion of plastic-elastic deformations, also increases the impact energy of the balls in the workpiece and thus translates into a greater increase in microhardness (Fig. 13). At the same time, as the speed increases v_g the thickness of the hardened layer g_h increases, the value of which is maximum 85 μm in relation to the core, which was verified by the statistical test regarding the average value.

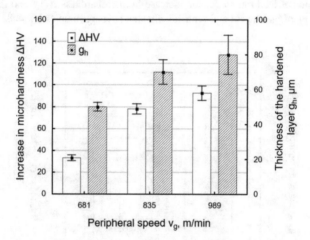

Fig. 13. Influence of peripheral velocity of the head v_g for the increase in microhardness ΔHV and thickness of the hardened layer g_h of plasma cut and centrifugally shot peening samples (v_f = 3648 mm/min, f_p= 0.08 mm, $i = 1$)

An increase in the feed rate v_f (Fig. 14) and transverse feed f_p (Fig. 15) causes that machining tracks appear at a greater distance, which results in uneven deformation of the centrifugally shot peening surface. The use of a low feed rate v_f or a low transverse feed value f_p causes that number of impacts per unit area is higher, which increases the thickness of the hardened layer g_h. In the range of feed rate $v_f = 3648 \div 5928$ mm/min and transverse feed $f_p = 0.08 \div 0.13$ mm the analyzed parameters have a greater impact on the thickness of the hardened layer g_h.

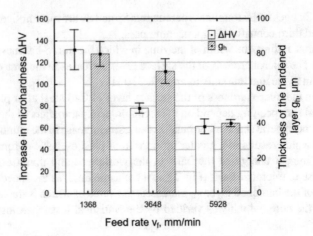

Fig. 14. Influence of feed rate v_f for the increase in microhardness ΔHV and thickness of the hardened layer g_h of plasma cut and centrifugally shot peening samples ($v_g = 835$ m/min, $f_p = 0.08$ mm, $i = 1$)

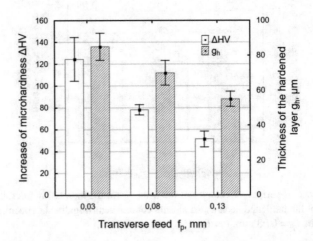

Fig. 15. Influence of transverse feed f_p for the increase in micro hardness ΔHV and thickness of the hardened layer g_h of plasma cut and centrifugally shot peening samples ($v_g = 835$ m/min $v_f = 3648$ mm/min, $i = 1$)

Figure 16 presents the impact of the number of centrifugal shot peening passes i on the increase in microhardness and thickness of the hardened layer. In the range of $i = 1 \div 2$ the microhardness of the surface layer increases as the number of passes increases. The use of the number of centrifugal shot peening passes larger than $i = 2$ results in a minimal increase in microhardness. This may be due to the phenomenon of "energy supersaturation".

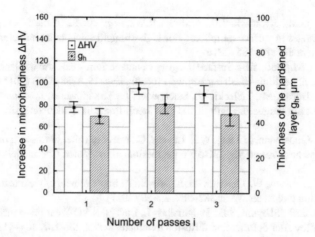

Fig. 16. Influence number of centrifugal shot peening passes i for the increase in microhardness ΔHV and thickness of the hardened layer g_h of plasma cut and centrifugally shot peening samples ($v_g = 835$ m/min, $v_f = 3648$ mm/min, $f_p = 0.08$ mm)

4 Conclusions

The following conclusions can be drawn based on the centrifugal shot peening researches of C45 steel parts cut with plasma:

1. As a result of centrifugal shot peening of plasma-cut C45 steel elements, the Sa, Sz, Sp, Sv parameters improved. After centrifugal shot peening, a maximum reduction of roughness parameters of up to 5.3 times was achieved.
2. An increase in the peripheral speed of the head v_g and the number of centrifugal shot peening passes reduces the roughness parameters, while an increase in the feed rate v_f and transverse feed f_p translates into higher values of the analyzed parameters.
3. Centrifugal shot peening of plasma cut elements resulted in a maximum increase in microhardness $\Delta HV = 132$, and the thickness of the reinforced layer is from 40 μm to 85 μm.
4. The increase in peripheral speed v_g and the number of passes i causes an increase in microhardness and thickness of the reinforced layer g_h. On the other hand, an increase in feed rate v_f and transverse feed f_p decreases the analyzed variables.
5. Based on the centrifugal shot peening researches carried out, it can be concluded that this method of machining can be successfully carried out on CNC machine tools.

References

1. Maawad, E., Bromeier, H.-G., Wagner, L., Sano, Y., Genzel, Ch.: Investigation on the surface SND near – surface characteristics of Ti-2.5Cu after various mechanical surface treatments. Surf. Coat. Technol. **205**, 3644–3650 (2011)

2. Krolczyk, G.M., Krolczyk, J.B., Maruda, R.W., Legutko, S., Tomaszewski, M.: Mertro-logical changes in surface morphology of high–strength steels in manufacturing processes. Measurement **88**, 176–185 (2016)
3. Salahshoor, M., Guo, Y.B.: Surface integrity of magnesium – calcium implants processed by synergistic dry cutting–finish burnishing. Procedia Eng. **19**, s288–s293 (2011)
4. Kubit, A., Bucior, M., Zielecki, W., Stachowicz, F.: The impact of heat treatment and shot peening on the fatigue strength of 51CrV4 steel. Procedia Struct. Integrity **2**, 3330–3336 (2016)
5. Zhang, P., Lindemann, J., Ding, W.J., Leyens, C.: effect of roller burnishing on fatigue prop-erties of the hot – rolled Mg-12Gd-3Y magnesium alloy. Mater. Chem. Phys. **124**, 835–840 (2010)
6. Toboła, D., Brostow, W., Czechowski, K., Rusek, P.: Improvement of wear resistance of some cold working tool steels. Wear **382–383**, 29–39 (2017)
7. Pohreyluk, I.M., Sheykin, S.E., Padgurskas, J., Lavrys, S.M.: Wear resistance of two-phase titanium alloy after deformation-diffusion treatment. Tribol. Int. **127**, 404–411 (2018)
8. Konefal, K., Korzynski, M., Byczkowska, Z., Korzynska, K.: Improved corrosion resistance of stainless steel X6CrNiMoTi17-12-2 by sidle diamond burnishing. J. Mater. Process. Technol. **213**, 1997–2004 (2013)
9. Menezes, M.R., Godoy, C., Buono, V.T.L., Schvartzman, M.M.M., Wilson, J.C.A.-B.: Effect of shot peening and treatment temperature on wear and corrosion resistance of sequentially plasma treated AISI316L steel. Surf. Coat. Technol. **309**, 651–662 (2017)
10. Kwiatkowski, M., Kłonica, M., Kuczmaszewski, J., Satho, S.: Comparative analysis of ener-getic properties of Ti6Al4 V Titanium and EN-AW-2017A(PA6) Aluminum alloy surface layers for an adhesive bonding application. Ozone Sci. Eng. **35**(3), 220–228 (2014)
11. Rudawska, A., Reszka, M., Warda, M., Miturska, I., Szabelski, J., Stancekova, D., Skoczylas, A.: Milling as a method of surface pre-treatment of steel for adhesive bonding. J. Adhes. Sci. Technol. **30**(23), 2619–2636 (2016)
12. Świrad, S., Wydrzynski, D., Nieslony, P., Krolczyk, G.M.: Influence of hydrostatic burnishing strategy on the surface topography of martensitic steel. Measurement **138**, 590–601 (2019)
13. Maheshwari, A.S., Gawande, R.R.: Influence of specially designed high-stiffness ball bur-nishing tool on surface quality of titanium alloy. Materials Today: Proc. **4**, 1405–1413 (2017)
14. Kułakowska, A., Kukiełka, L., Kukiełka, K., Malag, L., Patyk, R., Bohdal, L.: Possibilty of steering of products surface layer properties in burnishing rolling process. Appl. Mech. Mater. **474**, 442–447 (2014)
15. Zaleski, R., Zaleski, K., Gorgol, M., Wiertel, M.: Positron annihilation study of aluminium, titanium, and iron alloys surface after shot peening. Appl. Phys. A Mater. Sci. Process. **120**, 551–559 (2015)
16. Wiertel, M., Zaleski, K., Gorgol, M., Skoczylas, A., Zaleski, R.: Impact of impulse shot peening parameters on properties of stainless steel surface. Acta Phys. Pol., A **132**(5), 1611–1615 (2017)
17. Sahoo, B., Satpathy, R.K., Prasad, K., Ahmed, S., Kumar, V.: Effect of shot peening on low cycle fatigue life of compressor disc of a typical fighter class aero-engine. Procedia Eng. **55**, 144–148 (2013)
18. Gao, Y.-K., Lu, F., Yin, Y.-F., Yao, M.: Effects of shot peening on fatigue properties of 0Cr13Ni8Mo2Al steel. Mater. Sci. Technol. **19**, 372–374 (2003)
19. Galda, L., Sep, J., Prucnal, S.: The effect of dimples geometry in the sliding surface on the tribological properties under starved lubrication conditions. Tribol. Int. **99**, 77–84 (2016)
20. Zaleski, K.: The effect of vibratory and rotational shot peening and wear on fatigue life of steel. Eksploatacja i Niezawodnosc Maint. Reliab. **19**(1), 102–107 (2017)

21. Zebrowski, R., Walczak, M., Klepka, T., Pasierbiewicz, K.: Effect of the shot peening on surface properties of Ti-6Al-4V alloy produced by means of DMLS technology. Eksploatacja i Niezawodnosc Maint. Reliab. **21**(1), 46–53 (2019)
22. Matuszak, J., Zaleski, K.: Analysis of deburring effectiveness and surface layer properties around edges of workpieces made of 7075 aluminium alloy. Aircraft Eng. Aerosp. Technol. **90**(3), 515–523 (2018)
23. Skoczylas, A.: Geometric structure of the C45 steel surface after centrifugal burnishing and perpendicular shot peening. Adv. Sci. Technol. Res. J. **12**(2), 20–28 (2018)
24. Korzyński, M., Pacana, A.: Centreless burnishing and influence of its parameters on machining effects. J. Mater. Process. Technol. **210**, 1217–1223 (2010)
25. Nadolny, K., Plichta, J., Radowski, M.: Reciprocal internal cylindrical grinding integrated with dynamic centrifugal burnishing of hard-to-cut materials. Proc. Inst. Mech. Eng. Part E: J. Process Mech. Eng. **229**(4), 265–279 (2015)
26. Bhowmick, S., Basu, J., Majumdar, G., Bandyopadhyay, A.: Experimantal study of plasma arc cutting of AISI304 stainless steel. Mater. Today: Proc. **5**, 4541–4550 (2018)

Analysis of EDM Drilling of Small Diameter Holes

Marcin Płodzień⬤, Jarosław Tymczyszyn⬤, Witold Habrat$^{(\boxtimes)}$⬤,
and Piotr Kręcichwost

Faculty of Mechanical Engineering and Aeronautics, Rzeszow University of Technology,
Al. Powstancow Warszawy 12, 35-959 Rzeszow, Poland
`habrat@prz.edu.pl`

Abstract. In the paper, the EDM drilling using a brass tubular electrode was analyzed. The tests were carried out with the use of 42CrMo4 steel and the Accutex AH-35ZA EDM machine. Deionized water was used as the dielectric. The influence of feed speed and current intensity on eroding time, selected parameters of the surface roughness and wear of electrode were determined. The electrode wear was carried out directly on the machine by measurement of electrode contact with the top surface of the workpiece before and after machining. Macrogeometry and surface roughness parameters of the holes were analyzed based on the results of measurements by the ALICONA INFINITE FOCUS optical microscope. The tests showed that as the feed rate and current increase, the eroding time decreases. For higher current values in the adopted range, tool wear indicated higher intensity. A significant influence of feed and current on shape accuracy has been demonstrated. The lowest roughness parameters for the highest feed rate were obtained.

Keywords: EDM drilling · Electrode wear · Shape accuracy · Surface roughness

1 Introduction

Electrical Discharge Machining (EDM) is one of the machining methods where the material is removed as a result of a series of rapid electrical discharges in the area between the electrode and workpiece. The workspace is filled with a dielectric. The energy in the form of an electric discharge is converted into heat energy, causing melting and evaporation of the workpiece material. Material (workpiece and used electrode) is removed due to the dielectric flow. EDM machining is an alternative solution for forming elements made of hard materials with high strength such as carbides, composite materials, and ceramics. This machining makes it possible to make slender holes, where the diameter-to-depth ratio is significantly less than 1/20. During EDM drilling the electrode kinematics is like for standard drilling (Fig. 1), performing rotary and feed motion [1].

The physics of material removal in the EDM drilling process is complex. As a result of applying voltage to the electrodes, the dielectric ionizes in the inter-electrode gap. In the place where the distance between the electrode is the smallest, there is a decrease in the

© Springer Nature Switzerland AG 2020
G. M. Królczyk et al. (Eds.): IMM 2019, LNME, pp. 186–200, 2020.
https://doi.org/10.1007/978-3-030-49910-5_17

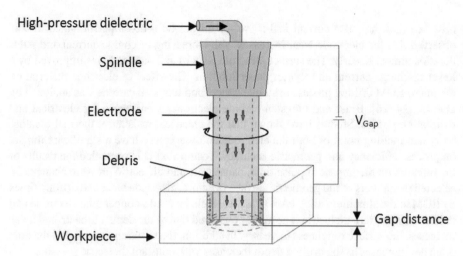

Fig. 1. Diagram of the electrical discharge drilling process

electrical strength of the medium, which results in an electric discharge. Holes made by the EDM micro-drilling method have a regular shape and high surface accuracy without burrs. However, the machining parameters that ensure the achievement of appropriate quality indicators have a significant impact on process efficiency and tool wear [2, 3].

Recently, miniaturized products and devices that are "smaller and cheaper" are increasingly being produced. EDM technology enables the production of complicated geometrical shapes and difficult-to-process materials of any hardness and strength, as long as it conducts electricity, e.g. tool steels and aviation alloys. Currently, micro EDM drilling has found wide application in industry. This applies to the production of molds and matrices, biomedicine, the automotive industry and aviation engineering. Typical examples include inkjet nozzles, spinning holes for synthetic fibers; cooling holes for aircraft turbine blades; injection nozzles in the automotive industry; dental and surgical implants, cooling ducts for cutting tools and fuel filters [4–10]. Many research works included process analysis and optimization of micro EDM drilling parameters. During EDM, there are several factors that directly affect the quality indicators of the process. Many factors can be associated with process parameters affecting performance (e.g. voltage, peak current) or system (e.g. type of dielectric fluid, tool properties, chemical and physical properties of the material). Many researchers are taking into account material removal rate and electrode wear rate, and relative wear in order to contribute to achieving high-quality product in a shorter time and at lower manufacturing costs [11, 12]. Li et al. [13] analyzed the effect of various dielectrics, differing in electrical conductivity, on the surface structure of the borehole by EDM in a nickel alloy. They found that as the electrical conductivity of the dielectric fluids increased, the transformation layer became thicker. The rate of cooling of the dielectric fluid affected the microstructure of the transformed layer, where the surface of the transformed layer was able to form a fine-grained zone. Ay et al. [12] worked on optimizing the EDM micro drilling process at Inconel 718. In order to determine the best conditions, they analyzed the parameters of the discharge current and pulse duration. Analyzing the optimization results, they observed that the pulse current had a greater impact on process performance. It was observed that the electrode wear increases with increasing discharge current and pulse duration almost linearly. The surface structure (cracks and damage) was improved by a lower discharge current and shorter pulse duration. The effect of electrode material on the micro EDM drilling process in stainless steel and tungsten carbide was analyzed by D et al. [14–16]. Brass and tungsten carbide electrodes were tested for electrical and thermal properties. Studies have shown that the electrical resistance, thermal conductivity and melting point of both the electrode and workpiece have a significant impact on process efficiency and geometric accuracy. Sosinowski [17] presented the results of the research on the impact of pulse time, pause time and dielectric pressure changes on selected parameters of the geometric surface structure, electrode wear and drilling times by EDM in the aluminum alloy PA6. In his research, he used a copper tube electrode and deionized water as a dielectric. The author observed that as the depth of the drilled holes increases, the surface roughness increases. In addition, the research showed that the hole diameter increases as the drilling depth increases with constant dielectric pressure.

Analysis of the state of the art in the field of EDM drilling showed the complexity of the issue and confirmed the need for further research in this area due to expectations from industry. The aim of the study was to analyze the impact of EDM micro-drilling machining parameters on selected quality indicators. The tests were carried out for a brass tubular electrode. The test was carried out using a constant electrode rotation speed, but with different ranges of process parameters in terms of current and feed rate of the tool.

2 Experiment Details

2.1 Experimental Research Conditions

Steel 42CrMo4 with a thickness of 10 mm was used for the tests. Table 1 shows the properties of the workpiece material. As a tool for EDM a tubular, brass electrode with an outer and inner diameter of 2 mm and 0.6 mm, respectively, was used (Table 2). Deionized water was used as the dielectric.

The Accutex AH-35ZA EDM machine was used for the tests (Fig. 2).

Table 1. Physical properties of workpiece materials

Physical Properties	42CrMo4
Density [g/cm^3]	7.80
Melting temperature [°C]	740–770
Electrical resistivity [Ωcm]	$8.2 \cdot 10^{-6}$
Thermal conductivity [W/mK]	40.0–45.0
Specific heat [J/(g°C)]	0.460–0.480

Source: www.matweb.com

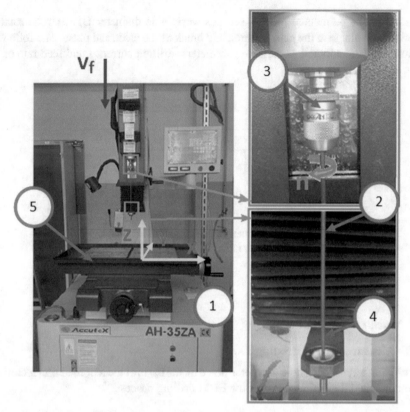

Fig. 2. Test stand: 1 – Accutex AH-35ZA EDM machine, 2 – working electrode, 3 – toolholder, 4 – pilot, 5 – work space

Table 2. Physical properties of electrode materials

Physical properties	Brass C26800
Density [g/cm^3]	8.47
Melting temperature [°C]	905–930
Electrical resistivity [Ωcm]	$6.63 \cdot 10^{-6}$
Thermal conductivity [W/mK]	121
Specific heat [J/(g°C)]	0.38

Source: www.matweb.com

2.2 Design of Experiment

The EDM drilling process was carried out on the basis of a full three-level design in accordance with the parameters presented in Table 3. For the tests, the technological and electrical parameters of the drilling process determined at the initial testing stage were used. The permanent process settings were: hole diameter D, electrode rotation speed n, the voltage in the gap, duration and break of the electrical pulse. The following parameters were adopted as variable parameters: drilling current I and feed rate of the electrode f.

Table 3. Machining parameters

Sample No.	1	2	3	4	5	6	7	8	9
D, mm n, obr/min					2 10				
I, A	26	26	26	28	28	28	31	31	31
f, mm/min	20	30	40	20	30	40	20	30	40

Legend: D - diameter of the tool, n - rotational speed, I - current intensity, f - feed

2.3   Measurement Methodology

2.3.1   Measurement of Eroding Time

The eroding time of each hole was determined on the basis of the recorded film recording, which included the entire course of the EDM drilling process.

2.3.2   Measurement of Eroding Wear

The measurement of the length of electrode wear was carried out directly on the Accutex AH-35ZA machine, on which holes were made. The measurement of consumption consisted in determining the difference between the initial and final length of the electrode. The electrode wear was measured by the operation of contact of the electrode with the workpiece before and after treatment. Figure 3 presents a graphic diagram of the measurement of electrode length wear.

2.3.3   Preparation of the Sample for Measurements

Before measuring of the macrogeometry and topography of the EDM holes, samples were properly prepared. The tested sample required intersection along the axis of symmetry of the holes. This process was carried out using the Mitsubishi FA10S EDM cutter, which is shown in Fig. 4.

2.3.4   Measurement of Macrogeometry and Roughness and Wave Parameters of the Surface of Holes Made by EDM Drilling

Measurement of dimensional and shape accuracy of holes was carried out with the use of the INFINITE FOCUS optical microscope from ALICONA. Measurements were made with an x5 magnification lens, and surface topography was measured with an x10 magnification lens (Fig. 5). The dimensional accuracy analysis consisted in measuring the radius r of holes in planes normal to the axis of the hole. The distance between measuring cross-sections was 0.5 mm on the length 10 mm (Fig. 6). Based on the cross-section profiles of the surface of the holes after drilling EDM, selected parameters of roughness (*Ra*, *Rz*) and wave of the tested surface (*Wa*, *Wz*) were determined.

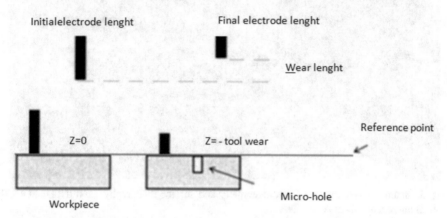

Fig. 3. Diagram of measuring consumption over the length of working electrodes

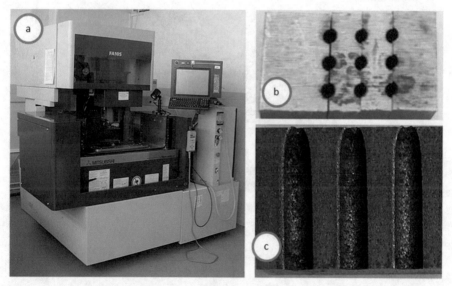

Fig. 4. Stand for sample preparation: a – MitsubishiFA10S EDM, b - cut test specimen along the axis of the holes, c - cross-sectional view of the holes made by the EDM method

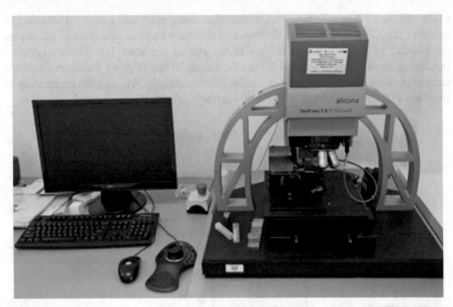

Fig. 5. Stand for assessment of macrogeometry and surface topography - INFINITE FOCUS optical microscope from ALICONA

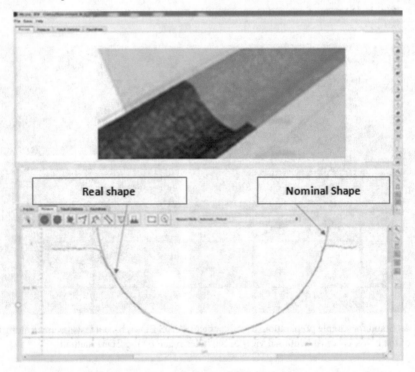

Fig. 6. Cross-sectional view of the scanned hole and shape accuracy analysis

3 Results and Discussion

3.1 EDM Drilling Time

Figure 7 shows the obtained values of hole machining time depending on the feed rate f and the current I during EDM drilling. It has been observed that as the feed speed f and the current I increase, the machining time decreases monotonically. It was shown that the shortest machining time took place at the highest current $I = 31$ A and the highest feed rate $f = 40$ mm/min and was about 25.7 s. It was observed that with increasing current and feed rate of electrode, machining efficiency increases. The m machining time for the highest value of feed rate from experimental range $f = 40$ mm/min decreases the process efficiency approximately 13% for $I = 26$ A, and about 9% for $I = 28$ A compared to the electrode operating at current $I = 31$ A. Moreover with a two-fold increase in the electrode feed rate for $I = 26$ A and $I = 31$ A, the machining time was reduced by about 8%.

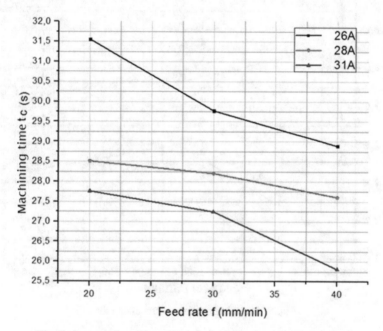

Fig. 7. Impact of working electrode feed speed on machining time

3.2 Electrode Wear

Figure 8 shows the results of electrode wear depending on the applied current I and feed rate f. Analyzing the obtained results, it can be seen that together with the increase of the feed rate f, a decrease in electrode wear was obtained for the value of current $I =$

28 A and 31 A, respectively. However, for current $I = 26$ A, an increase in electrode consumption was observed with an increase of the feed rate.

In the case of changes in the feed value, the lowest value of electrode wear was obtained for the current $I = 26$ A and the feed speed $f = 20$ mm/min, it was about 7.1 mm. Analyzing the values of electrode consumption for changes in current I, an increase in current was observed with an increase in current. The highest value of electrode wear was recorded for current $I = 28$ A and feed speed $f = 20$ mm/min, which was about 8.2 mm. For the current $I = 31$ A, for the feed speed $f = 30$ mm/min and $f = 40$ mm/min, the electrode wear value was similar and amounts to 7.9 mm. It can be observed that, depending on the adopted process conditions, the electrode consumption ranged from 70 to 82% of the drill hole depth.

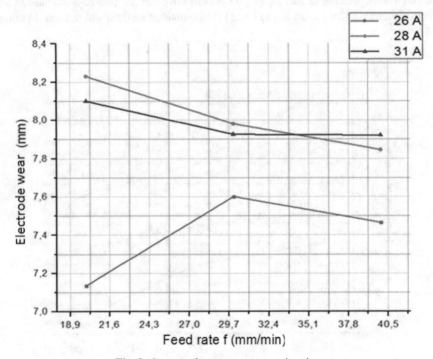

Fig. 8. Impact of current on processing time

3.3 Shape Accuracy

Figure 9 presents deviation maps showing the change in hole radius in relation to the distance from the top surface. The analysis of the obtained deviation maps shows that the largest dimensional and shape accuracy of the hole was obtained at the parameters $f = 40$ mm/min and $I = 31$ A. The cross-sections of the holes drilled by the EDM method resemble a barrel shape, at a hole depth of 2.5 mm to 6.5 mm recorded radius values were the largest. This may be due to worse conditions for flushing residual contaminants from the borehole and poorer cooling at greater depths. As a result, the electrical conditions in the gap are disturbed, causing short-circuit currents or premature electrical discharges.

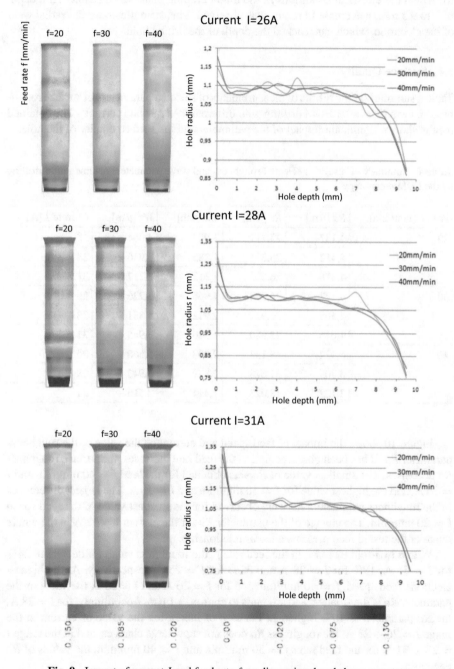

Fig. 9. Impact of current *I* and feed rate *f* on dimensional and shape accuracy

When analyzing the quality of hole on the top surface, an increase in radius in relation to the nominal radius can be observed. This condition is caused by the start of the electrode for which the electrical conditions in the initial eroding phase are unstable. At a depth of 8 to 9.5 mm, a decrease in radius was observed. This state illustrates the radial wear of the electrode, which has reached the depth of the drilled hole.

3.4 Surface Quality

Table 4 summarizes the results of measurements of selected parameters of surface roughness and waviness after EDM drilling with different feed rate and currents. The obtained results show a significant impact of the parameters on the surface quality of the hole.

Table 4. Summary of results of selected roughness and wave parameters obtained after drilling by the EDM technology

Feed f [mm/min]	Ra [μm]	Rz [μm]	Wa [μm]	Wz [μm]	Current I [A]
20	4,132	23,711	2,007	4,064	26
	4,412	26,251	1,285	3,936	28
	4,478	26,003	1,313	3,217	31
30	4,177	24,594	1,966	5,226	26
	4,505	26,912	2,912	5,381	28
	4,046	22,880	1,520	3,945	31
40	3,122	18,133	2,008	5,560	26
	3,709	21,678	2,879	3,945	28
	3,122	18,028	1,414	3,516	31

Figure 10 shows the impact of feed speed and current on the value of the roughness parameter Ra. It has been observed that as the feed rate increases, the surface roughness Ra decreases. The smallest value of Ra was obtained for the feed $f = 40$ mm/min and $I = 26$ A, and the highest value for $f = 30$ mm/min and $I = 28$ A. The largest differences in the roughness parameters Ra were observed for the smallest value of the feed speed $f = 20$ mm/min. The change of the parameter Ra for the current I = 31 A in the whole range of the tested feed presents a linear relationship.

With a two-fold increase in the feed rate f, the roughness value Ra decreased 24% for $I = 31$ A, 18% for $I = 28$ A and 32% for $I = 26$ A, respectively. At the highest electrode feed speed, $f = 40$ mm/min and for $I = 26$ A and $I = 31$ A the value of the parameter Ra is the smallest and amounts to approx. 3.1 μm. Accordingly, for $I = 28$ A, the Ra parameter is 19% higher. In the case of increasing the value of current in the range $I = 26 \div 28$ A, the roughness Ra does not show large changes, and in the range $I = 28 \div 31$ A for the feed rates $f = 30$ mm/min and $f = 40$ mm/min, the values of Ra decreases.

Figure 11 shows the dependence of the Rz roughness parameter on the feed rate and current. The dependence of the Rz parameter on the process parameters is similar in nature to the roughness parameter Ra.

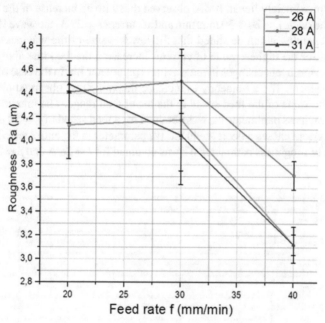

Fig. 10. Dependence of the surface roughness parameter Ra of the drilled hole by EDM method on the feed rate f and current I.

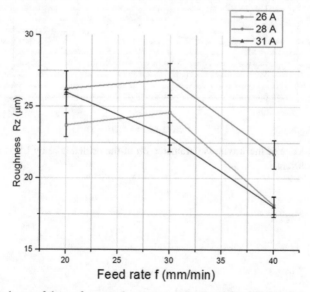

Fig. 11. Dependence of the surface roughness parameter Rz of the drilled hole by EDM method on the feed rate f and current I.

Graphical results of measurements of *Wa* and *Wz* surface waviness parameters of holes made by EDM method are shown in Figs. 12, 13. The smallest value of *Wa* parameter was achieved at a feed rate $f = 20$ mm/min and current $I = 28$ A. Dependence of the waviness parameter *Wa* on the feed rate for electrodes operating at $I = 26$ A and $I = 31$ A is approximately linear. It was observed that with an increase of the value of the feed rate in the range $f = 20 \div 30$ mm/min and at current $I = 28$ A, the wave *Wa* increased by 125%. Generally, it can be stated that the lowest value of the waviness parameters was obtained for the highest value of current. Increasing the feed rate above 30 m/min did not cause significant changes in the waviness parameter *Wa*. In the case of the surface waviness parameter *Wz*, a tendency to increase the value was observed along with an increase in the value of the feed speed in the lower feed range. The maximum value of the surface waviness parameter *Wz* was obtained for a feed speed of $f = 30$ mm/min. An increase in current causes a decrease in the *Wz* parameter in the entire tested range. The lowest value of roughness parameter *Wz* was obtained for $I = 28$ A and $f = 20$ mm/min.

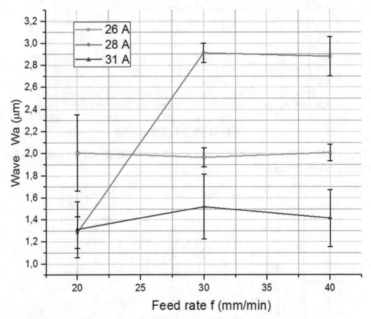

Fig. 12. Dependence of the waviness parameter *Wa* of the drilled hole by EDM method on the feed rate f and current I

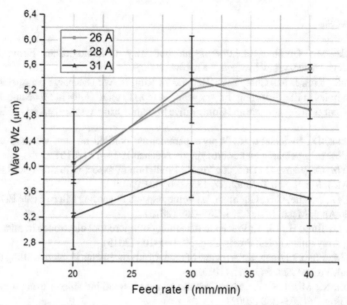

Fig. 13. Dependence of the waviness parameter Wz of the drilled hole by EDM method on the feed rate f and current I

4   Conclusions

Based on the research on the impact of current and electrode feed rate on machining time, electrode wear, the accuracy of hole diameter, surface roughness parameters Ra and Rz and waviness parameters Wa, Wz, the following conclusions can be drawn:

- an increase in the feed rate reduces the machining time,
- the greatest longitudinal wear of the working electrode occurs at current $I = 28$ A for the lowest feed rate value $f = 20$ mm/min,
- on the top surface, the large increase of hole diameter relative to the nominal radius was observed, which may be due to a place of sudden electrical discharges,
- diameter accuracy of EDM holes significantly depends on the measurement depth of the hole,
- it is difficult to clearly determine the dependence of roughness parameters and waviness parameters on feed rate. The lowest roughness parameters for the highest feed rate were obtained.

References

1. Siwczyk, M.: Obróbka elektroerozyjna podstawy technologiczne. Firma Naukowo-Techniczna "Mieczysław Siwczyk," Kraków (2000)
2. Lee, C.-S., Kim, J.-M., Choi, Y.-C., Heo, E.-Y., Kim, D.-W.: Discharge parameters for pass-through holes in EDM-drilling. In: Azevedo, A. (ed.) Lecture Notes in Mechanical Engineering. Advances in Sustainable and Competitive Manufacturing Systems. pp. 617–626. Springer (2013)
3. Oniszczuk, D., Świerszcz, R.: Wpływ parametrów obróbki na stan warstwy wierzchniej elementów po wycinaniu elektroerozyjnym. Mechanik. 14–17 (2015)
4. Li, J., Yin, G., Wang, C., Guo, X., Yu, Z.: Prediction of aspect ratio of a micro hole drilled by EDM. J. Mech. Sci. Technol. **27**, 185–190 (2013)
5. Dodun, O., Panaite, C.E., Gherman, L.: Some Aspects Regarding Micro-Scale EDM Drilling Process. Appl. Mech. Mater. **371**, 285–289 (2013)
6. Zhang, L., Tong, H., Li, Y.: Precision machining of micro tool electrodes in micro EDM for drilling array micro holes. Precis. Eng. **39**, 100–106 (2015)
7. Hasan, M., Zhao, J., Jiang, Z.: A review of modern advancements in micro drilling techniques. J. Manuf. Process. **29**, 343–375 (2017)
8. Egashira, K., Mizutani, K.: Micro-drilling of monocrystalline silicon using a cutting tool. Precis. Eng. **26**, 263–268 (2002)
9. Horiuchi, O., Masuda, M., Shibata, T.: Bending of drill and radial forces in micro drilling. In: Advanced Materials Research, pp. 642–648 (2013)
10. Fu, W.K., He, N., Li, L., Zhao, M., Bian, R.: Research on micromechanical drilling of micro-hole array in PVC mask. In: Materials Science Forum, pp. 239–243 (2014)
11. Yilmaz, V., Sarikaya, M., Dilipak, H.: Deep micro-hole drilling for hadfield steel by electro-discharge machining (EDM). Mater. Tehnol. **49**, 377–386 (2015)
12. Ay, M., Çaydaş, U., Hasçalik, A.: Optimization of micro-EDM drilling of inconel 718 superalloy. Int. J. Adv. Manuf. Technol. **66**, 1015–1023 (2013)
13. Li, C., Xu, X., Li, Y., Tong, H., Ding, S., Kong, Q., Zhao, L., Ding, J.: Effects of dielectric fluids on surface integrity for the recast layer in high speed EDM drilling of nickel alloy. J. Alloys Compd. **783**, 95–102 (2019)
14. D'Urso, G., Maccarini, G., Ravasio, C.: Process performance of micro-EDM drilling of stainless steel. Int. J. Adv. Manuf. Technol. **72**, 1287–1298 (2014)
15. D'Urso, G., Maccarini, G., Ravasio, C.: Influence of electrode material in micro-EDM drilling of stainless steel and tungsten carbide. Int. J. Adv. Manuf. Technol. **85**, 2013–2025 (2016)
16. D'Urso, G., Giardini, C., Ravasio, C.: Effects of electrode and workpiece materials on the sustainability of micro-EDM drilling process. Int. J. Precis. Eng. Manuf. **19**, 1727–1734 (2018)
17. Sosinowski, Ł.: Elektroerozyjne wiercenie otworów o małych średnicach w materiałach o dużej przewodności cieplnej. Mechnik. 9–14 (2015)

The Influence of Traverse Speed on Geometry After Abrasive Waterjet Machining

Maciej Bartkowiak[1], Michał Wieczorowski[2], Natalia Swojak[2(✉)], and Bartosz Gapiński[2]

[1] Volkswagen Poznan, ul. Warszawska 349, Poznan, Poland
[2] Institute of Mechanical Technology, Division of Metrology and Measurement Systems, Poznan University of Technology, ul. Piotrowo 3, 60-965 Poznan, Poland
natalia.swojak@put.poznan.pl

Abstract. The paper discusses findings presenting the effect of traverse speed on the surface quality, obtained due to the procedure of processing with the use of abrasive waterjet (AWJ). The analyses of surface deviations were carried out for the samples of three different materials, i.e. steel S235JR, aluminium alloy PA6 and PE WUHD 1000, applied to sample cuttings within the range of traverse speeds from 25% up to 200% by using a 25% gradient. The traverse speed corresponding to 100% was factory-defined for each tested material. Barton garnet was used as an abrasive material; mesh 80. The study was conducted with the maintenance of constant water jet pressure and the abrasive mass flow rate of 350 MPa and 148 g/min, respectively. According to the guidelines of EN ISO 9013:2017-04 standard, the analysis of two parameters characterizing a cutting-site groove shape was made, namely thickness diminution (Δa) and deviation from perpendicularity or inclination (u). Hence, surface deviation values were found in terms of macrogeometry. Based on the conducted studies, the effect of the AWJ processing speed values on the geometry of a produced groove was defined and depicted as a trendline. Also, we have drawn conclusions on the possibilities to find an optimum machining speed for a given material, which would enable us to obtain the most favourable final geometry. The authors are therefore convinced that it is possible to obtain favourable surface finish of semi-products after the AWJ machining.

Keywords: Abrasive waterjet machining (AWJ) · Cutting · Surface quality · Traverse speed

1 Introduction

Before further processing, the majority of materials need to be prepared. Very frequently, they need to undergo cutting. At the contemporary stage of technological development, materials can be applied to cutting by using a variety of machines. In the practice, the cutting can produce waste (using a turning lathe; frame-, disk-, band- or abrasive cutters, oxygen and plasma ones) or be waste-free (using scissors or impact-related). Apart from

© Springer Nature Switzerland AG 2020
G. M. Królczyk et al. (Eds.): IMM 2019, LNME, pp. 201–213, 2020.
https://doi.org/10.1007/978-3-030-49910-5_18

the above mentioned commonly used methods of cutting, there are plenty of other methods considered as unconventional, which have been more and more popular in industrial plants. This is the case especially in the war, aircraft and electronic industries when it comes to the manufacture of press tools, templates, mold cavities or dams, where special almost unprocessable materials, products made of special ceramics and composites are used [1–3]. This principle also refers to elements of inconsiderable dimensions, of small roughness, as well as products of demanding properties as to their superficial layer.

Main unconventional methods of material cutting include the following: abrasive waterjet (AWJ) machining, cutting with a string armoured with an abradant, laser processing, anode-mechanical cutting, and electroerosive machining (cutting out).

2   Abrasive Waterjet (AWJ) Machining

Waterjet was first used as a tool in ore mining in 1940. The advantages of waterjet lead to the development of this technology and an increase in its use in industrial practice. On one hand, it allows shaping very soft materials (rubber, fiber composite materials, foam materials, cellulose mass etc.), on the other – very hard ones [4] (steel, metals and non-iron alloys, alloys, ceramics, glass, titan).

The rule of developing jet consists in using water to speed up abrasive grains (most frequently garnet).

After being mixed with water in a mixing chamber and formed in a so-called mixing tube, the abradant forms a water-abradant jet of a specific diameter [5] capable of cutting hardest materials [6]. The phenomenon of the so-called ejector abradant suction is taking place. A very big velocity of water under big pressure causes the underpressure below the nozzle of the mixing tube [7]. This results in an automatic suction of the abradant to the chamber, where it is mixed with water. The water-abradant mixture is then formed in the mixing tube and directed into the machining site. An important benefit of AWJ machining is little effect of strains on the element that is being machined. It gives enormous possibilities of manufacturing parts whose well-thickness is of about 0.5 mm with no concerns that these could break. An extremely positive phenomenon is the fact that it is the force of the jet itself that acts on the machined material, suppressing it to the grid. Special plastic pads which protect the part that is being cut out against falling on the bottom of the bath may be used while cutting out very small elements.

In the literature on the subject matter, an effect of AWJ machining on surface roughness [8–10] has been investigated, especially with reference to the machining velocity [11]. Spatial parameters connected with surface topography [12] have been used in the roughness assessment. The analysis is being carried out also with regard to emerging vibrations [13] and the re-usage of the material [14]. Taking into consideration varying reliability and comparability of the measurements' results [15], different metrology techniques, both contact and contactless [16], can be used to analyze surface roughness on the microscale. Given that, it seems very interesting and justifiable to make an approach aimed at predicting surface quality obtained as a result of AWJ machining while particular process parameters are taken into account [17].

There are fewer publications in the literature on changes in the geometry on the macroscale [18]. Defining the effect of machine cutting speed on surface macrogeometry was therefore the major motivational factor of the conducted studies. The AWJ is known as a machining method utilizable for the cutting of almost all types of materials without taking risk of any thermal deformations [9]. However, having made a literature review we failed to find any comparative studies on the effect of machine cutting speed on surface macrogeometry in the case of materials considerably distinct as to their physicochemical properties. Our article also provides readers with a valuable knowledge from the field of technological process development depending on a processed material, with a simultaneous production of satisfactory surface macrogeometry.

Various metrology techniques may be used for macroscale measurements. Most frequent are optical measurements which use optical scanners, measurement microscopes, and recently also computed tomography (CT) [19]. In our preliminary work, we attempted to use optical scanning techniques, taking into account their usefulness on both macro- and meso-scales [20, 21].

3 Materials and Methods

ATMS Water Jet TK-1010-FA machine has been chosen for the experiment. It is a three-section workstation: big pressure section, the system of mechanic head movement, and the system of programming and steering of the machine. From the point of view regarding the cutting process accuracy aspects, its mechanic part plays a pivotal role. Its construction consists in an open boom system mounted on a stationary basis, on which the moving of perpendicularly situated beams against one another takes place. Next, an axis of each beam corresponds to the Cartesian system axis, using the right hand rule. A digital drive coupled with a ball screw enables repositioning in a given axis. The sole reposition, however, occurs on the profiled linear slide bearings. This section of the station is presented in Fig. 1.

Fig. 1. Extension arm system used to move the cutting head [22].

A cutting head installed on the test machine is an IDE head. The head construction enables a manual change of the tilt angle against the plane of the machined material (in the 'Z' and 'X' planes of the machine). The machine used in the experimental work is designed to cut diverse materials. Therefore, three different material samples of the same height were used in our study. Material blocks of the following dimensions were used in our experiment: length 400 mm, width 150 mm and height 35 mm. The samples were preliminarily prepared by means of the chip cutting method using milling. Materials for the samples were as follows: S235JR (St3S) Steel, PA 6 aluminium alloy and PE WUHD 1000 polyethylene.

To the best of our knowledge, in order to make an in-depth analysis of the feed rate effect on cutting quality, for each studied sample we have decided to perform seven test cuts within the range from 25% up to 200%, with increments every 25%. The baseline speed (100%) was defined in factory for each of the materials as the content of abrasive, due to the optimization of the cutting process. The parameter subjected to change was the feed rate. For each of the speed rates the abrasive flow rate remained unchanged at 148 g/min. The constant process parameters are shown in Table 1.

Table 1. Constant parameters and their values.

Constant parameters	Water jet pressure [MPa]	Abrasive mass flow rate [g/min]	Focusing tube diameter [mm]	Standoff [mm]	Abrasive type/ Abrasive size
Value	350	148	1,2	3	Barton Garnet/ 80 mesh

The appearance of samples after the cutting is shown in the following figures: steel S235JR (Fig. 2), aluminium alloy PA6 (Fig. 3), PE WUHD 1000 (Fig. 4).

Fig. 2. Sample S235JR.

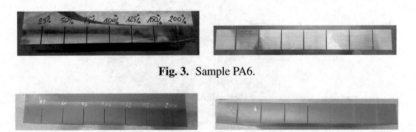

Fig. 3. Sample PA6.

Fig. 4. Sample PE WUHD 1000.

To evaluate the quality of shaping by cutting, including abrasive waterjet cutting, the following standard is used: EN ISO 9013: 2017-04 Thermal cutting - Classification of thermal cuts - Geometrical product specification and quality tolerances. In order to unequivocally assess the results of the conducted research, methodology included in this standard was used, and in accordance with its content results were determined.

According to the scheme of the carried out research on the groove shape in the case of perpendicular cutting for AWJ machining, characteristic values are the following quantities (Fig. 5):

- Δa - thickness diminution;
- u - deviation from perpendicularity or inclination.

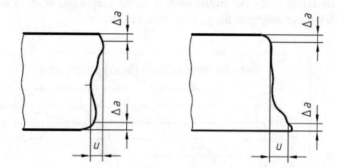

Fig. 5. Thickness diminution and deviation from perpendicularity or inclination (vertical cutting) [23].

Based on these characteristic values, the analysis of the cutting head feed rate effect on the groove geometry was carried out. The measurements were made on the station equipped with Stemi 508 microscope coupled with a digital camera and dedicated software. Methodology of the analysis was shown in Fig. 6.

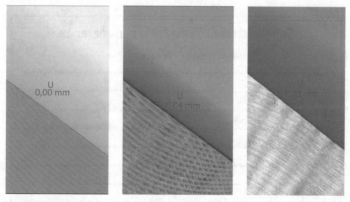

Fig. 6. The methodology of sample analysis and values corresponding to the speed at the level of 25% of the assumed speed.

4 Results and Discussion

On the basis of the performed analyses, the results were obtained, which are shown below in a tabular and graphic form in order to highlight the trend of changes during changes of the traverse speed for the tested material. The values Δa are presented separately for the top and bottom part of the sample.

For the sample made of S235JR material measurement results (Table 1) and the trend of the deviation of the cut profile walls from the plumb line (Fig. 7) and the edge loss on the input and output of the jet (Fig. 8) are as follows:

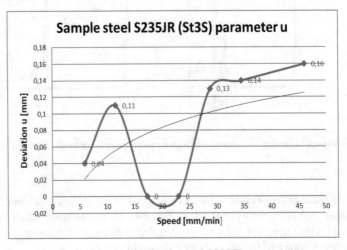

Fig. 7. Analysis results for the steel S235JR – parameter u.

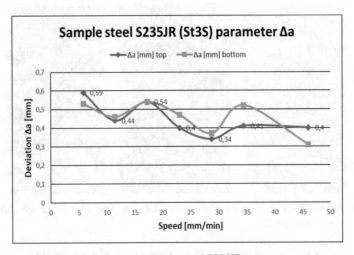

Fig. 8. Analysis results for the steel S235JR – parameter Δa.

For the sample made of PA6 measurement results (Table 2) and the trend of the deviation of the cut profile walls from the plumb line (Fig. 9) and edge loss on the input and output of the jet (Fig. 10) are as follows:

Table 2. Analysis results for the sample made of steel S235JR.

Sample No.	Speed [mm/min]	Speed [%]	Δa [mm]		u [mm]
			Top	Bottom	
1	5,8	25	0,59	0,53	0,04
2	11,5	50	0,44	0,46	0,11
3	17,3	75	0,54	0,54	0
4	23	100	0,4	0,47	0
5	28,8	125	0,34	0,37	0,13
6	34,5	150	0,41	0,52	0,14
7	46	200	0,4	0,31	0,16

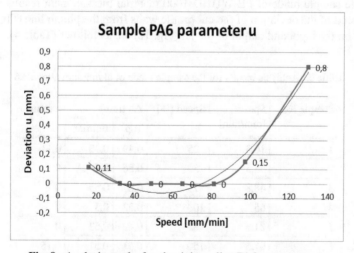

Fig. 9. Analysis results for aluminium alloy PA6 – parameter u.

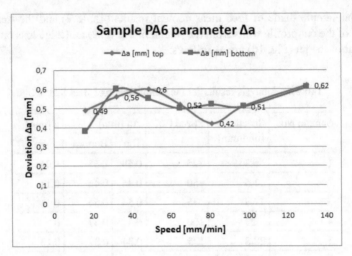

Fig. 10. Analysis results for aluminium alloy PA6 – parameter Δa.

For the sample made of PE WUHD 1000 material measurement results (Table 3) and the trend of the deviation of the cut profile walls from the plumb line (Fig. 11) and edge loss on the input and output of the jet (Fig. 12) are as follows (Table 4):

Table 3. Analysis results for the sample made of aluminium alloy PA6.

Sample No.	Speed [mm/min]	Speed [%]	Δa [mm] Top	Bottom	u [mm]
1	16,3	25	0,49	0,38	0,11
2	32,5	50	0,56	0,6	0
3	48,8	75	0,6	0,55	0
4	65	100	0,52	0,5	0
5	81,3	125	0,42	0,52	0
6	97,5	150	0,51	0,51	0,15
7	130	200	0,62	0,61	0,8

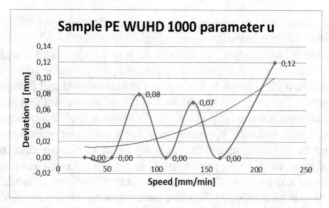

Fig. 11. Analysis results for polypropylene PE WUHD 1000 – parameter u.

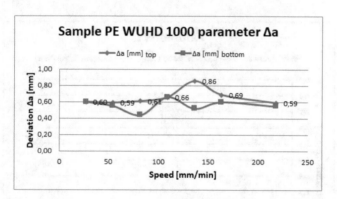

Fig. 12. Analysis results for polypropylene PE WUHD 1000 – parameter Δa.

Table 4. Analysis results for the sample made of polypropylene PE WUHD 1000.

Sample No.	Speed [mm/min]	Speed [%]	Δa [mm]		u [mm]
			Top	Bottom	
1	27,3	25	0,6	0,6	0
2	54,5	50	0,59	0,55	0
3	81,8	75	0,61	0,44	0,08
4	109	100	0,66	0,66	0
5	136,3	125	0,86	0,52	0,07
6	163,5	150	0,69	0,60	0
7	218	200	0,59	0,55	0,12

Data analysis made on the basis of a series of test cuttings gives a preview of the trend occurring along with increasing the working feed rate during the AWJ cutting. Collecting (Fig. 13) the carried out analyses for each of the tested materials on one chart, clearly demonstrates that along with a growing cutting rate, the deviation of the cut material wall perpendicularity in relation to the axis of the abrasive waterjet increases, referred to as parameter u. The clear fact should be also taken into account that at low traverse speeds a similar phenomenon takes place, too. However, it is of decreasing nature along with growing speed, until the optimum speed for a particular workpiece is reached. After crossing this optimum speed, the geometry of the groove deteriorates in relation to the axis of abrasive waterjet. For materials of a higher density this course is more easily visible and unequivocal. Comparing the obtained results one should expect that such a scheme will be duplicated for other materials available for the AWJ treatment, while bearing in mind that in an analogous manner it will reflect the working feed rate parameter with respect to the mechanical properties of workpiece.

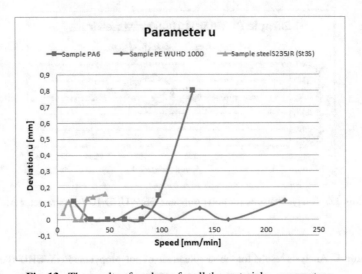

Fig. 13. The results of analyses for all the materials – parameter u.

Performed tests with regard to the deformation and loss of material at the edges of the grooves from the input and output of the jet also demonstrate a relationship between the influence of the working feed rate and the deformation for a particular type of material (Fig. 14), but it is no longer so evident as in the case of the perpendicularity of the walls with respect to the axis of the jet. However, it can be observed that for this parameter (Δa) there is also such a traverse speed, at which the loss of the edge of material or deformation reach the minimum values. Analysing the two parameters, a relationship between the deviation from the axis of the cut material wall, and the loss or deformation of the cut material's edge can be easily seen. Namely, for the two measured parameters there is such a speed for which the deviation is the smallest. Furthermore, this speed for both the parameters is present within the same range, making it the optimum speed in the AWJ machining process for a given material. Due to a continuous development of the technology described here, it should be presumed that the greater the pressure the increase of machining speed for workpiece material will be possible, while maintaining or even increasing current accuracy values.

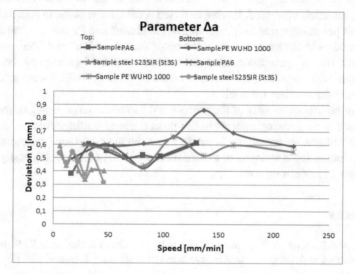

Fig. 14. The results of analyses for all the materials – parameter Δa.

5 Conclusions

Abrasive jet machining is getting more and more popular mean of cutting various materials. For the macro-scale evaluation of geometry in the post-cutting section and for the review of superficial sites of unevenness (in micro-scale), a variety of measuring devices, operating based on contact or contactless methods, can be used. The studies presented in this article demonstrate our view on the possibilities of using this specific machining in industrial applications and also for atypical materials.

Our study indicates, among others, that it is possible to determine an optimum range of the traverse speed, at which the final geometry of a processed material would be the best. In the case of steel the minimum values of deviations for the "u" parameter exist at the speed of approximately 20 min/min. A similar tendency is manifested by "Δa" parameter, for which the minimum deviations can be observed within the speed range of 20–30 mm/min. Steel presents greater Δa deviations at smaller speeds, and smaller Δa deviations at greater speeds. In the case of aluminium the scope of speeds at which the "u" parameter deviations are the smallest, is broader than in the case of steel, and equals from 35 up to 80 mm/min. Also, the effect of speed on "Δa" parameter changes, where smaller deviations are found at lower speeds, while greater deviations are observed at higher speeds. Optimum values of the speed for "Δa" parameter are within the range of 65–80 mm/min, at the same time being equivalent to the speed values at which it is feasible to achieve the smallest deviations of the "u" parameter.

Bearing in mind the polypropylene (PP), it is possible to determine several values of the speed, at which the "u" parameter deviations are the smallest, their values being 50, 110 and 160 mm/min, respectively. In the case of the PP the effect of speed on the change of values of "Δa" parameter deviations is the weakest compared to the remaining materials discussed. Having analysed accumulative plots, in comparison with steel, it was perceivable that the processing of aluminium and PP could be carried out at greater speeds with the maintenance of the optimum value of "u" and "Δa" parameters. Importantly, the "u" parameter value dramatically increased as soon as the speed of 100 mm/min was exceeded. The greatest values of "Δa" deviations were achieved for the PP, while the smallest for steel.

Due to the irreplaceability of the process and its numerous advantages, the method will be used with growing interest in the procedures of cutting. Another reason for its greater and greater popularity is the fact that the material-based technology experiences an increasing development towards composite materials, which easily undergo the abrasive jet machining nowadays.

References

1. Kovacevic, R., Hashish, M., Mohan, R., Ramulu, M., Kim, T.J. Geskin, E.S.: State of the art of research and development in Abrasive Waterjet Machining, J. Manuf. Sci. Eng, 119(4B), 776–785
2. Youssef, H.A., Hassan El-Hofy, H.: Machining Technology. Machine Tools and Operations. CRC Press, Boca Raton (2008)
3. Miller, D.S.: Micromachining with abrasive waterjets. J. Mater. Process. Technol. 149(1–3), 37–42 (2004)
4. Cárach, J., Hloch, S., Hlaváček, P., Ščučka, J., Martinec, P., Petrů, J., Zlámal, T., Zeleňák, M., Monka, P., Lehocká, D., Krolczyk, J.: Tangential turning of Incoloy alloy 925 using abrasive water jet technology. Int. J. Adv. Manuf. Technol. 82(9–12), 1747–1752 (2015)
5. Orbanić, H., Junkar, M., Bajsić, I., Lebar, A.: An instrument for measuring abrasive water jet diameter. Int. J. Mach. Tools Manuf 49(11), 843–849 (2009)
6. Kinik, D., Gánovská, B., Hloch, S., Monka, P., Monková, K., Hutyrová, Z.: On-line monitoring of technological process of material abrasive water jet cutting. Tehnički vjesnik 22(2), 351–357 (2015)
7. Hreha, P., Radvanská, A., Hloch, S., Peržel, V., Królczyk, G., Monková, K.: Determination of vibration frequency depending on abrasive mass flow rate during abrasive water jet cutting. Int. J. Adv. Manuf. Technol. 77(1–4), 763–774 (2014)
8. Kovacevic, R.: Surface texture in abrasive waterjet cutting. J. Manuf. Syst. 10(1), 32–40 (1991)
9. Begic-Hajdarevic, D., Cekic, A., Mehmedovic, M., Djelmic, A.: Experimental study on surface roughness in abrasive water jet cutting. Procedia Eng. 100, 394–399 (2015)
10. Hashish, M.: Characteristics of surfaces machined with abrasive-Waterjets, J. Eng. Mater. Technol. 113(3), 354–362
11. Löschner, P., Jarosz, K., Niesłony, P.: Investigation of the effect of cutting speed on surface quality in abrasive water jet cutting of 316L stainless steel. Procedia Eng. 149, 276–282 (2016)
12. Klichova, D., Klich, J., Zlamal, T.: The use of areal parameters for the analysis of the surface machined using the abrasive waterjet technology. In: Hloch, S., Klichova, D., Królczyk, G.M., Chattopadhyaya, S. (eds.) Advances in Manufacturing Engineering and Materials. Springer, Heidelberg (2018)

The Influence of Traverse Speed 213

13. Hreha, P., Radvanska, A., Knapcikova, L., Królczyk, G.M., Legutko, S., Królczyk, J.B., Hloch, S., Monka, P.: Roughness parameters calculation by means of on-line vibration monitoring emerging from AWJ interaction with material. Metrol. Measur. Syst. XXII **2**, 315–326 (2015)

14. Schnakovszky, C., Herghelegiu, E., Radu, M.C., Tampu, N.C.: The surface quality of AWJ cut parts as a function of abrasive material reusing rate. In: Modern Technologies in Industrial Engineering, ModTech2015, IOP Publishing, IOP Conference on Series: Materials Science and Engineering, vol. 95, 012004 (2015)

15. Wieczorowski, M., Cellary, A., Majchrowski, R.: The analysis of credibility and reproducibility of surface roughness measurement results. Wear **269**(5–6), 480–484 (2010)

16. Mathia, T.G., Pawlus, P., Wieczorowski, M.: Recent trends in surface metrology. Wear **271**(3–4), 494–508 (2011)

17. Hloch, S., Valíček, J.: Topographical anomaly on surfaces created by abrasive waterjet. Int. J. Adv. Manuf. Technol. **59**(5–8), 593–604 (2012)

18. Kong, M.C., Axinte, D.: Response of titanium aluminide alloy to abrasive waterjet cutting: geometrical accuracy and surface integrity issues versus process parameters. Proc. Inst. Mech. Eng. Part B: J. Eng. Manuf. **223**(1), 19–42 (2009)

19. Gapinski, B., Wieczorowski, M., Marciniak-Podsadna, L., Dybala, B., Ziolkowski, G.: Comparison of different methods of measurement geometry using CMM, optical scanner and computed tomography 3D. Procedia Eng. **69**, 255–262 (2014)

20. Gapiński, B., Wieczorowski, M., Marciniak-Podsadna, L., Swojak, N., Mendak, M., Kucharski, D., Szelewski, M., Krawczyk, A.: Use of white light and laser 3d scanners for measurement of mesoscale surface asperities, In: Diering, M., Wieczorowski, M., Brown, C.A. (eds.) rozdział w: Advances in Manufacturing II, V.5, Metrology and Measurement Systems, pp. 239–256. Springer (2019)

21. Majchrowski, R., Grzelka, M., Wieczorowski, M., Sadowski, L., Gapiński, B.: Large area concrete surface topography measurements using optical 3D scanner. Metrol. Measur. Syst. **XXII**(4), 565–576 (2015)

22. http://atmsolutions.pl/

23. EN ISO 9013:2017-04 Thermal cutting – Classification of thermal cuts - Geometrical product specification and quality tolerances

Analysis of Hole Positioning Accuracy with the Use of Position Deviation Modifiers

Dawid Wydrzyński[✉], Jacek Bernaczek, Grzegorz Budzik, Marek Magdziak, and Grzegorz Janas

Department of Mechanical Engineering, Rzeszow University of Technology, Al. Powstańców Warszawy 12, 35–959 Rzeszów, Poland
dwydrzynski@prz.edu.pl

Abstract. The development of serial production has created a need for normalisation of technical documentation for the acceptable range of dimensional variability and product geometry. The introduction of uniform rules reduced production costs and reduced the number of products rejected at the quality control stage. The aim of the paper is to presents the use of the principle of maximum material requirements (MMR) for functional tolerance. The principle of maximum and minimum material, as well as the principle of reciprocity, are still not sufficiently used in industrial applications. Accordingly, appropriate research models were designed, working drawings were made without the use of a modifier, as well as with the use of a position modifier to demonstrate the validity of applying the principle of maximum material in relation to assembly tolerances. The developed models and working drawings present the construction and technological requirements that must be met in order for this principle to be applied. Then mock–up models were made on the Haas VF2 CNC milling machine. Functional models were measured using a measuring machine. An analysis of the obtained results is presented. On its basis, it was determined which models were made correctly and where the position deviation modifier proved to work. The correct application of the maximum material requirement, least material requirement and reciprocity requirement allows clearly describing the functional properties of dimensioned elements specified by the constructors using the largest possible tolerances. This gives great economic benefits for producers. However, it is very important that the functionality of the set and the satisfactory cooperation of elements can be achieved with many combinations of dimensional and geometric deviations. Functional tolerance of the cumulative effect of dimensional and geometric deviations (shape, direction and position) gives very significant technical and economic benefits in a market with price and quality competition.

Keywords: Position deviation · Minimum material requirement · Maximum material requirement · Reciprocity requirement

1 Introduction

The development of serial production has created a need for normalisation of technical documentation for the acceptable range of dimensional variability and product geometry. The introduction of uniform rules reduced production costs and reduced the number

© Springer Nature Switzerland AG 2020
G. M. Królczyk et al. (Eds.): IMM 2019, LNME, pp. 214–225, 2020.
https://doi.org/10.1007/978-3-030-49910-5_19

of products rejected at the quality control stage. Therefore, there is a strong need to pay more attention to tolerance design to enable high precision sets to be produced at lower costs. In the automotive and aviation industries, the tolerance process is a significant problem in the design and manufacture of products. It is important to define a consistent expression of the geometric product specification during the tolerance process of the whole life cycle. Hence, it is necessary to integrate the individual stages of product design and production so that it is possible to quickly manage data on the changes [1–3]. Dimensional and geometrical accuracy, as well as in the final stage, the assembly and cooperation of elements is very important from the point of product functionality. As the technology advances and performance requirements constantly tighten, the costs and required precision of the sets will increase. Also in modern technologies, i.e. rapid prototyping techniques, there is a strong need for tolerance for the production of parts with increased accuracy [4–6]. Correct use of tolerance makes communication between the designer and technologist complete and makes misinterpretation impossible. Geometric tolerances are used above normal dimensional tolerances when it is necessary to more accurately control the shape or location of the characteristics of the manufactured part, due to the nature of the work that the part must perform. Geometric tolerances are used to concisely and accurately convey the full geometric requirements in technical drawings. They should always be taken into account for surfaces that come into contact with other parts, especially when strict tolerances are used for given elements [7, 8]. Engineers can use modern CAx engineering programs which can assist in their work, however, these systems do not always have the solutions that are accordance with the standards [9–11].

When checking the geometry of objects, dimensions and geometric deviations are measured separately. To avoid economic losses (rejection of products during quality control) maximum material requirement (MMR), least material requirement (LMR) and reciprocity requirement (RPR) was introduced in the ISO 2692:2008 [12] and ASME Y14.9 standards [13]. Maximum material requirement (MMR) and least material requirement (LMR) are well established for functional tolerance and are used to guarantee assembly options. Their use gives significant technical and economic benefits [14, 15] as well as ecological, because the smaller number of parts rejected as a result of quality control, reduces the consumption of resources which otherwise would adversely affect the environment [16].

These principles have been described in the above standards and have been applied since the 90s of the last century [17, 18]. The ability to apply these requirements allows clearly describing the constructor's requirements with the largest possible tolerances [19]. To facilitate the understanding of the functioning and application of the principles of tolerance with the use of position modifiers, Humienna Z. and Turek P. present the use of animation techniques as a tool to explain the complex concept of maximum material requirements (MMR) [20]. Anselmetti B. and Laurent P. proposed two complementary assumptions and some explanations regarding the concepts used in the ISO 2962: 2014 standard so that the definitions comply with the definitions of the ISO 1101: 2012 standard [21]. Laurent P. et al. demonstrate the LMR's ability to develop requirements for loose assemblies and conformity assessment in dimensional metrology based on the virtual size of the smallest material [22]. In turn, Tang Z. states that the simultaneous application of maximum material requirements in terms of reference tolerance and concentricity of shaft parts can ensure assembly and reduce costs. Existing DMMR concentricity assessments use either inflexible real functional indices or slow mathematical methods, which limits the use of such good tolerance in industry [23]. This article presents the use of the principle of maximum material requirements (MMR) for functional tolerance. The principle of maximum and minimum material, as well as the principle of reciprocity, are still not sufficiently used in industrial applications. Accordingly, appropriate research models were prepared to demonstrate the validity of applying the principle of maximum material in relation to assembly tolerances. The developed models and working drawings present the construction and technological requirements that must be met in order for this principle to be applied. The case study developed as part of the work shows the significant benefits of using MMR in economic and technological terms.

2   Materials and Methods

The research model assumes joining of two plates milled with base pins. Two types of plates were assumed. First with holes: ⌀16H7 and ⌀12H7 for the base pins. Roundness and hole position tolerances were introduced (Fig. 1).

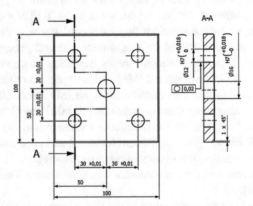

Fig. 1. Working drawing of the plate without the use of modifiers.

The second model assumes the implementation using the so–called maximum material modifier. For clarity and functionality, bases A and B have been introduced on two walls of the tile, to which the maximum material requirement will apply. An additional third base was placed on the axis of one of the holes, which will be the base hole. The position of the holes was determined using theoretically accurate dimensions contained in the frames. The axis of the hole should be exactly in the place designated by the constructor. The maximum material requirement was applied to the axis of the base hole. The value of the hole diameter should, therefore, be ⌀12 mm + 0.3 mm. This tolerance is much wider than assumed in the first model of the plate ⌀12H7(+0.018) mm. This is due to the possibility of non–compliance with the roundness tolerance requirement for the hole (0.02 mm) and perpendicularity of the hole axis (0.02 mm) to the base, this is appropriate because the plate could be rejected at the quality control stage even though it could turn out to be assembled and functional. To additionally emphasize the sense of using the modifier, the axis of the base hole during manufacture will be intentionally shifted by 30 ± 0.01 mm in the horizontal direction. The working drawing of the part is presented below (Fig. 2).

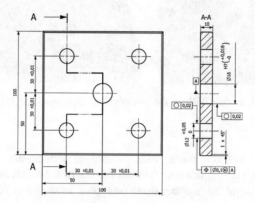

Fig. 2. Working drawing of the plate using the maximum material modifier.

The study assumes that two plates will be joined together during construction, for which a modifier has not been used. Then the model with the maximum material requirement will be connected to the plate made without a modifier. The image of the set made in the NX 11.0 program containing both variants of milled plates, as well as base pins is shown in the figure below (Fig. 3).

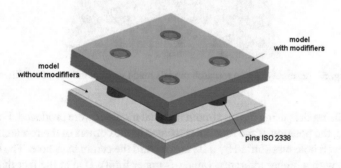

Fig. 3. The set consists of two milled inserts and base pins.

The plates are made of C40 steel with dimensions of $100 \times 100 \times 10$ mm. The semi–finished product was ground to obtain a smooth, flat surface and then transferred to a station equipped with a numerically controlled milling machine. The processing was performed on a Haas VF–2 milling machine. After performing the above operations, a research model of the plate was obtained without using the maximum material modifier. The effect of the processing is shown below in Fig. 4a. After the processing, the base pins were installed to the base model, as shown in Fig. 4b below.

a)   b)

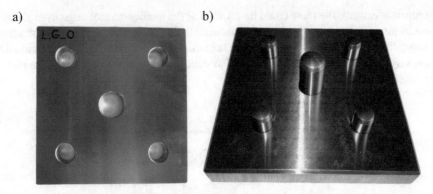

Fig. 4. Base research model. Without the position deviation modifiers (a) and with fitted base pins (b).

A second plate was also made, with dimensions which were consistent with the assumptions of the working drawing without the use of a maximum material modifier. Both plates were assembled together using base pins, as shown in Fig. 5 below.

Fig. 5. Assembly of two research models made without the use of modifiers.

Next, the models using the maximum material modifier were produced. For its use to make sense, the position of the four holes located in the corners of the research plate was changed. Each hole was shifted by 0.02 mm toward the centre base hole. The holes have been made with a higher tolerance value (its upper limit). Due to the fact that the plate processing program with the use of the modifier was analogous, it was changed manually on the machine tool panel. The four outer holes were reamed with a ⌀12.04 mm reamer. These values are selected in such a way that both tiles can be assembled to each other. This is a key issue for the functionality of the parts as well as the tested set. The effect of the work on the first model with the modifier is shown below in Fig. 6a. Next, both tiles are assembled together, which shows that the second part, which did not meet the appropriate dimensional tolerances regarding the position of the holes, is mountable by making holes in the upper tolerance deviation (Fig. 6b).

a)   b)

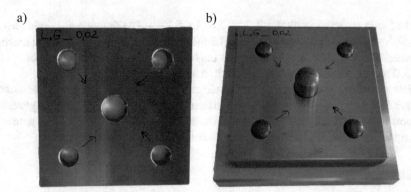

Fig. 6. Model using the modifiers. Model with shifted hole axes toward the centre of the plate (a) and the assembled set (b).

Next, a model was made with a larger offset of 0.04 mm towards the centre base hole axis. The size of the holes was also increased. These were made using a ⌀12,06 mm size reamer. The resultant model, thanks to the use of a modifier and making the holes in the upper tolerance limits, despite the change in the position of the holes, was functional and mountable in a set with the plate which dimensions were correct. In order to verify the accuracy of individual dimensions of the research models, they were measured on a coordinate measuring machine. The plates were measured on a Carl Zeiss Accura II machine. The measuring range of the machine in three axes is respectively: X 1200 mm, Y 1800 mm, Z 1000 mm, and accuracy: $\pm 1.2 + L/350$ μm.

In the next stage of research, a measurement plan was made for all the model features defined in it. Measurements of individual features were made on the basis of the prepared drawing (Fig. 7), where on the "top" view of the research model the appropriate numbers indicate the order of position measurements (dimensions from 1 to 8) of individual holes

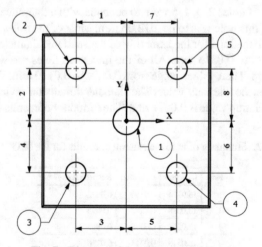

Fig. 7. Diagram of the measurement plan.

(marked with numbers in circles 2, 3, 4, 5) (Fig. 5.1.4.). The drawing also indicates the X and Y axes of the coordinate system established in the coordinate measuring machine software. The Y–axis was determined on the basis of the designated centres of holes 2 and 3. The measurement was carried out in manual mode.

The measurement process for all tested models was carried out in an analogous manner. For each plate, the program used to the coordinate measuring machine generated a report presenting the results of the measurement for each of the verified features. It contained the nominal value, upper and lower tolerance value and deviation.

3   Results

The first tested model was made nominally, without deliberate displacement of the holes and without using the maximum material requirement (marked in the pictures as "LG0"). In this case the dimension that exceeds the tolerance is the measurement of the position of hole No. 3 in the Y–axis (dimension of position No. 4). The deviation is too large by 0.0004 mm. However, as it turned out in practice, this value is so small that it does not interfere with the correct assembly of the tile set. Other values of the position of the hole axis are within the tolerance given in the working drawing. Their deviations range from −0.0048 to 0.0012 mm. The list of dimension values and their deviation is presented in Table 1.

Table 1. List of position dimensions and deviations for the "LG" model.

Dimension	1	2	3	4	5	6	7	8
Dimension value	−30,0022	29,9984	−30,0012	−30,0104	30,0018	−30,0048	29,9968	30,0017
Deviation	−0,0022	−0,0016	−0,0012	−0,0104	0,0018	−0,0048	−0,0032	0,0017

The diameters of holes 2, 3, 4, 5 were to be made with a tolerance of +0.03 mm, i.e. the upper value of this hole was to be ⌀12.03 mm. Most of the holes are slightly larger, which is however beneficial for the assembly of the set. Hole diameter deviations are in the range of 0.0288 to 0.0313 mm. All of the measured holes are within the assumed roundness deviation. Their values range from 0.0092 to 0.0110 mm. The diameter of the centre hole is within the tolerance value. The diameter dimension deviation is 0.0154 mm. The roundness deviation value is 0.0072 mm. The summary of results is shown in Table 2 below.

Table 2. Summary of hole measurement results for the "LG" sample.

Hole	Diameter	Deviation	Roundness deviation
1	16,0154	0,0154	0,0072
2	12,0288	0,0288	0,0098
3	12,0306	0,0306	0,0092
4	12,0313	0,0313	0,0110
5	12,0311	0,0311	0,0131

The second report refers to the plate that was also made without using position deviation modifiers. The positions of the holes in the X and Y axes have been in 6 out of 8 cases kept within tolerance boundaries. The value of the axis position Y of hole number 5 was slightly exceeded by 0.0009 mm, and the value of the position of hole number 2 on the Y–axis by 0.0022 mm. The hole position deviations range from 0.0001 mm to 0.0122 mm. The summary of results is shown in Table 3 below.

Table 3. Summary of position dimensions and deviations for the model "LG0".

Dimension	1	2	3	4	5	6	7	8
Dimension value	−29,9999	30,0122	−29,9995	−29,9978	30,0026	−29,9969	30,0035	30,0109
Deviation	0,0001	0,0122	0,0005	0,0022	0,0026	0,0031	0,0035	0,0109

Three out of four diameters were made correctly. The dimension deviations range from 0.0289 to 0.0317 mm. All of the roundness deviations are within the tolerance with values from 0.0060 to 0.0139 mm. Table 4 presents a summary of the results.

Table 4. Summary of hole measurement results for the sample "LG0".

Hole	Diameter	Deviation	Roundness deviation
1	16,0131	0,0131	0,0060
2	12,0305	0,0305	0,0135
3	12,0317	0,0317	0,0139
4	12,0313	0,0313	0,0092
5	12,0289	0,0311	0,0104

Measurement report number 3 contains information about the plate where the holes in the corners of the milled plate were deliberately moved during its manufacture by 0.02 mm in the X and Y axes in the direction of the central base hole axis. Therefore, only two dimensions regarding the position of the hole are within the assumed tolerance on the working drawing. Interestingly, you can see here the error of execution by a CNC milling machine, or an error when entering data when editing the program on the machine's panel. The dimensional deviations of the position of the holes in the X and Y axes are from 0.0069 to 0.0251 mm. The summary of results is shown in Table 5 below.

Table 5. List of position dimensions and deviations for the model "LG002".

Dimension	1	2	3	4	5	6	7	8
Dimension value	−29,9771	29,9951	−29,9749	−29,9931	29,9849	−29,9881	29,9838	29,9864
Deviation	0,0229	−0,0049	0,0251	0,0069	−0,0151	0,0119	−0,0162	−0,0136

Hole diameters 2, 3, 4, 5 were made in accordance with the maximum material requirement. Their diameter could, therefore, be a maximum of 12.05 mm. This depends on the basic hole tolerance, which is +0.05 mm. The measured deviations range from 0.0452 mm to 0.0462 mm. Thus, all four holes are within the tolerance due to the application of maximum material requirement, and also affect the functionality of the product, which, if not for use of the modifier, would probably be rejected due to failure to maintain the position of the holes. All roundness deviations are within the set tolerance and range from 0.0069 to 0.0189 mm. Hole position tolerances $\varnothing0.1$ mm have also been preserved and range from 0.0409 to 0.533 mm. The summary of results is shown in Table 6 below.

Table 6. List of hole measurement results for the sample "LG002".

Hole	Dimension	Deviation	Roundness deviation	Position deviation
1	16,0126	0,0126	0,0105	
2	12,0462	0,0462	0,0069	0,0409
3	12,0462	0,0462	0,0108	0,0533
4	12,0458	0,0458	0,0117	0,0467
5	12,0452	0,0452	0,0189	0,0421

The last measurement report describes the verification of the last research model, in which the axis of the holes in the corners was moved by 0.04 mm in the X and Y axes towards the axis of the central base hole. All of the 8 dimensions responsible for the position of the holes in the corners were made outside of the tolerance value. Their deviations range from 0.0277 mm to 0.0408. The results are shown in Table 7.

Table 7. List of position dimensions and deviations for the model "LG004".

Dimension	1	2	3	4	5	6	7	8
Dimension value	−29,9606	29,9723	−29,9592	−29,9618	29,9614	−29,9604	29,9629	29,9720
Deviation	0,0394	0,0277	0,0408	0,0382	0,0386	0,0396	−0,0371	−0,0280

The dimensions of the diameters of the holes in the corners were made outside the tolerance. During manufacture, they were milled in order to enlarge their diameters, so that the test set could be assembled. The deviations of the diameters of individual holes are therefore in a range of 0.0941 to 0.0949. When it comes to roundness deviations, only one of the holes is out of tolerance. This is hole number 2. The deviations range from 0.0091 to 0.0208 mm. Regarding the tolerance of individual tolerance positions, they fall within the tolerance after applying the maximum material modifier. The deviations range from 0.0879 to 0.1221 mm. The last test model should be rejected at the level of quality control as it would not be functional and mountable. In the case of these tests, it was assembled due to the enlargement of the holes, whose diameters, unfortunately, are outside the assumed tolerance. The list of results is illustrated in Table 8.

Table 8. List of hole measurement results for the sample "LG004".

Hole	Diameter	Deviation	Roundness deviation	Position deviation
1	16,0115	0,0115	0,0161	
2	12,0945	0,0945	0,0208	0,0879
3	12,0941	0,0941	0,0121	0,1182
4	12,0944	0,0944	0,0091	0,1221
5	12,0949	0,0949	0,0131	0,0912

4 Conclusion

Based on the tests carried out in this study, it should be stated that the modifiers can in a sense protect parts from rejection during quality control, when they meet functional requirements, and some of their tolerances are not met. However, this usually happens to a certain error value, after exceeding this the part must be disqualified at the stage of quality control. The advantages of using modifiers are obvious, although at the moment they are mainly used in the aviation and automotive industries.

– Parts no. 1 and 2 were fully functional because all of their important features from the assembly point of view were made correctly and their dimensions were within the tolerances given by the constructor.
– Research model number 3, whose positions of the tested holes were intentionally moved, was functional and mountable, even though the hole positions were out of tolerance. The application of the maximum material requirement, which conditioned the execution of holes in the upper limit of the tolerance value, resulted in the compliance of the part with the specification assumed in the working drawing.
– In turn, the last sample, despite the use of a modifier, could not be considered compatible, because its holes exceeded the upper value of the hole diameter deviation.
– The correct application of the maximum, least material and reciprocity requirements allows unambiguously describing the functional properties of dimensioned elements specified by the constructors using the largest possible tolerances. This gives great economic benefits for producers. However, it is very important that the functionality of the set; satisfactory cooperation of elements can be achieved with many combinations of dimensional and geometric deviations.
– Functional tolerance of the cumulative effect of dimensional and geometric deviations (shape, direction and position) gives very significant technical and economic benefits

in a market with price and quality competition. The advantages of using modifiers are obvious, although at the moment they are mainly used in the aviation and automotive industries.

References

1. Dantan, J.Y., Ballu, A., Mathieu, L.: Geometrical product specifications — model for product life cycle. Comput.-Aided Des. **40**, 493–501 (2008)
2. Desrochers, A., Laperriere, L.: Framework proposal for a modular approach of tolerancing. In: Procedia of the 7th CIRP International Seminar on Computer Aided Tolerancing, pp. 93–102 (2001)
3. Feng, S.C, Song, E.Y.: Information modeling of conceptual design integrated with process planning. In: Procedia of ASME DETC/DFM (2000)
4. Ameta, G., Lipman, R., Moylan, S., Witherell, P.: Investigating the role of geometric dimensioning and tolerancing in additive manufacturing. Trans. ASME: J. Mech. Des. **137**(11) (2015)
5. Moroni, G., Petro, S., Polini, W.: Geometrical product specification and verification in additive manufacturing. CIRP Ann.-Manuf. Technol. **66**, 157–160 (2017)
6. Leach, R.K., Bourellb, D., Carmignato, S., Donmez, A., Senina, N., Dewulf, W.: Geometrical metrology for metal additive manufacturing. CIRP Ann.–Manuf. Technol. **68**, 677–700 (2019)
7. Colin, H., Simmons, I., Dennis, E., Maguire, C., Neil, P.: Geometrical tolerancing and datums. In: Manual of Engineering Drawing, 3rd edn, pp. 159–167 (2009)
8. Ballu, A., Mathieu, L., Dantan, J.Y.: Global view of geometrical specifications. In: Bourdet, P., Mathieu, L. (eds.) Geometric Product Specification and Verification: Integration of Functionality. Springer, Dordrecht (2003)
9. Fengxia, Z., Kunpeng, Z., Linna, Z., Peng, Z.: Procedia research on the intelligent annotation technology of geometrical tolerance based on geometrical product specification (GPS). In: 13th CIRP Conference on Computer Aided Tolerancing, vol. 27, pp. 254–259 (2015)
10. Gaunet, D.: 3D functional tolerancing & annotation: CATIA tools for geometrical product specification. In: Bourdet, P., Mathieu, L. (eds.) Geometric Product Specification and Verification: Integration of Functionality. Springer, Dordrecht (2003)
11. Ngoi, B.K.A., Seow, M.S.: Tolerance control for dimensional and geometrical specifications. Int. J. Adv. Manuf. Technol. **11**, 34–42 (1996)
12. Smith, G.T.: Machining and monitoring strategies. In: Cutting Tool Technology. Springer, London (2008)
13. ASME Y14.5 2009. Dimensioning and tolerancing. The American Society of Mechanical Engineers, New York (2009)
14. Osanna, H., Tamre, M., Weckenmann, A., Blunt, L., Jakubiec, W.: In: Humienny, Z., (ed.) Geometrical Product Specifications–Course for Technical Universities, Warsaw University of Technology Printing House, Warsaw (2001)
15. Henzold, G.: Geometrical Dimensioning, and Tolerancing for Design, Manufacturing and Inspection. Butterworth-Heinemann, Oxford (2006)
16. Królczyk, G.M., Maruda, R.W., Królczyk, J.B., Wojciechowski, S., Mia, M., Nieslony, P., Budzik, G.: Ecological trends in machining as a key factor in sustainable production–a review. J. Clean. Prod. **218**, 601–615 (2019)
17. Jayaraman, R., Srinivasan, V.: Geometric tolerancing: virtual boundary requirements. IBM J. Res. Dev. **33–2**, 90–104 (1989)

18. Etesami, F.: Position tolerance verification using simulated gauging. Int. J. Robot. Res. **10–4**, 358–370 (1994)
19. Weckenmann, A., Werner, T.: Computer–assisted generation of individual training concepts for advanced education in manufacturing metrology. Measur. Sci. Technol. **21**, 379–383 (2010)
20. Humienny, Z., Turek, P.: Animated visualization of the maximum material requirement. Measurement **45**, 2283–2287 (2012)
21. Anselmetii, B., Laurent, P.: Complementary writing of maximum and least material requirements, with an extension to complex surfaces. In: 14th CIRP Conference on Computer Aided Tolerancing (CAT), vol. 43, pp. 220–225 (2016)
22. Laurent, P., Anselmetii, B., Nabil, A.: On the usage of least material requirement for functional tolerancing. In: 15th CIRP Conference on Computer Aided Tolerancing–CIRP CAT 2018, vol. 75, pp. 179–184 (2018)
23. Tang, Z., Huang, M., Sun, Y., Zhong, Y., Qin, Y.: Rapid evaluation of coaxiality of shaft parts based on double maximum material requirements. Measurement **147**, 1–13 (2019)

Impact of Tool Imbalance on Surface Quality in Al7075–T6 Alloy Machining

Dawid Wydrzyński[✉], Łukasz Przeszłowski, Grzegorz Budzik, and Bartosz Kamiński

Department of Mechanical Engineering, Rzeszow University of Technology,
Al. Powstańców Warszawy 12, 35–959 Rzeszów, Poland
dwydrzynski@prz.edu.pl

Abstract. The development of industry forces technologists to satisfy a continuous increase in the productivity and accuracy of manufactured elements. This is particularly evident in the processing of aircraft parts, where up to 95% of the semi–finished product is removed in machining. High–performance machining requires the use of tools with low runout and low imbalance. The unbalance itself is caused by the incorrect distribution of the rotating body mass in relation to its axis of rotation. Tool imbalance increases tool wear thus increasing degradation of machine tool components. These are not the only undesirable effects of tool imbalance, the poor surface quality after high–speed machining is also a factor in the importance of this issue. Measurements of the quality of surfaces machined with a tool with varying overhang and degree of unbalance showed that the unbalance values of the heat shrink holders used had little effect on SGP parameters, for small overhangs, obtained differences in parameter values were independent of unbalance. Surfaces made with tools with higher overhangs were characterized by higher roughness and waviness parameters, artificially introduced imbalance was also more important for them, which is particularly visible when comparing the wave profile of a surface machined with a balanced holder and the holder with the largest imbalance. Despite this, the unbalance values that heat shrink holders have after mounting the tool, without balancing, do not significantly affect the roughness parameters. The largest changes in the parameters of the geometric structure were recorded for the artificially introduced unbalance at the largest overhangs. It can be assumed that at higher spindle speed values, the obtained roughness index values would be higher due to the increase in the centrifugal force acting on the unbalanced mass, which is why balancing of tools is particularly important in the context of HSC machining or thin wall machining.

Keywords: Aluminium alloy milling · Tool imbalance · AA7075 · Tool runout

1 Introduction

In aviation constructions, aluminium alloys of the 2XXX series [1] and 7XXX [2, 3] are one of the most commonly used materials. The main advantage of these alloys is its high strength at a relatively low weight [4]. These properties are important in the aerospace industry because they reduce the weight of the aircraft structure and thereby reducing

© Springer Nature Switzerland AG 2020
G. M. Królczyk et al. (Eds.): IMM 2019, LNME, pp. 226–235, 2020.
https://doi.org/10.1007/978-3-030-49910-5_20

operating costs. At the same time, the development of industry forces technologists to satisfy a continuous increase in the productivity and accuracy of manufactured elements. This is particularly evident in the processing of aircraft parts, where up to 95% of the semi–finished product is removed in machining. High–performance machining requires the use of tools with low runout and low imbalance. Tool runout causes a number of adverse phenomena, which can include, among others deterioration of surface quality and tool wear [5–7].

The role of the tool holders in milling operations is similar to other machining operations, as high rigidity, clamping reliability and tool positioning accuracy is required. Both the tool mounting method and the type of tool affect the imbalance of the entire system. There are several types of unbalance, we focus mainly on so–called dynamic imbalance, which increases by a square of rotational speed. The standard specifies the requirements and calculates the allowable static and dynamic residual offset for rotary tools and tool systems [8–10]. Holder balancing is a key factor in high–speed machining. System imbalance depends on the unbalanced mass and its centre of gravity around the axis of rotation. Since it is impossible to achieve a perfectly balanced system, the goal is to reduce the unbalance to a minimum [11]. Very long tools, necessary for deep–hole pocketing, are very likely to cause vibration, loss of accuracy, as well as uneven and premature wear of the cutter. In such cases a balanced operation should be performed [12, 13]. The increasing use of high–speed machining (HSM) causes challenges for machine tool manufacturers and users. Simulation of cutting processes has become key to optimizing production: accurate forecasting of machining effects is now needed to improve milling performance. In the case of high performance or high cutting speeds, special attention should be paid to the quality of the machine–tool interface with regard to unbalance, stiffness and damping [14]. HSCM is widely used in the aviation industry for the production of structural construction elements [15, 16].

Weinert K. et al. [17] analysed the impact of runout using a dynamic milling process simulation. They compared the results with experimental data. One aspect is the difference in vibration with and without runout. In addition, a time series analysis method is presented to distinguish between vibration and runout. Due to different hollows of the blade, the surface structure is also produced in the process of doubling the period. The only way to detect runout during the process is through tool vibration analysis.

Franco P. et al. [18] analyse the impact of radial and axial runout on surface roughness during face milling with round insert cutting tools. They carry out milling tests to verify the correctness of the model. The discrepancies between the experimental and theoretical surface profile are a consequence of various factors, such as changes in undeformed chip thickness along the surface profile.

Yue C. et al. [19] summarize the current state of knowledge in the field of research on problems with chattering forecasting, their identification and control, damping, with particular emphasis on milling processes. Particular emphasis is placed on the theoretical relationship between cutting and damping of the process, tool runout and gyroscopic effect, as well as the importance of this phenomenon in forecasting vibrations.

Żabiński et al. [20] present the possibilities of using artificial intelligence in the process of detecting the unbalance of milling heads used in numerically controlled milling centres. They tested heads on four levels of unbalance. The data set comprised

27334 records with 14 features in time and frequency domain. They have experimentally proved that different degrees of unbalance of the milling head can be detected with high accuracy using computational intelligence methods. Detection of tool head unbalance should be used in the monitoring procedure of the machine performed after installing a new tool or after a specified period of its work.

Aquiar et al. [21] showed the dependence of the surface roughness of the object, tool wear and tool vibration when milling H13 steel using a slim tool. In the tests carried out, the roughness of the treated surfaces in most of their experiments was similar. In one experiment, where the tool had a large overhang, the surface roughness was much higher than in other studies. Based on the analysis of the force, the authors showed that in the tests with less roughness, the force signals showed different values at the same speed, which indicates the tools unbalance.

There are several methods for balancing tools. One of the methods was proposed by Alieva R. and Guseinov R., [22]. High–speed milling performance with long tool can be increased by supercritical speed machining. The method is based on self–alignment and self–cantering of the milling cutter. Based on the theoretical analysis of dynamic deformation, they developed a new milling cutter whose stem contains an internal chamber. Compensating material is introduced into the chamber to minimize residual unbalance. After passing through the critical zone, the long cutter works stably at high speed. There are also methods for balancing tools based on mass balancing. These methods are implemented on dedicated device [23]. One of the manufacturers of such equipment is the company Haimer. The device has the ability to determine the unbalance of the tool assembly and its correction masses in one (static balancing) or two correction planes (dynamic balancing).

The article presents the results of tests and measurements of the quality of surfaces machined with a tool of varying overhang and degree of unbalance. Milling tests were made using heat shrink tool holders of various lengths and cutters with diameters Ø10 and Ø16. The small number of publications available in scientific journals has become the basis for conducting this research.

2   Materials and Methods

The unbalance test was conducted on the HAIMER Tool Dynamic device. The aim of the study was to process samples with a balanced and unbalanced tool for fixed values of machining parameters, and then to measure the parameters of the geometrical structure of the treated surfaces. Milling tests were carried out using heat shrink tool

Table 1. Tools used.

Nr	Name	Diameter [mm]	Total length [mm]	Working length [mm]	Number of teeth	Fixed overhang [mm]
1	End mill cutter Ø10–22	10	72	22	3	35
2	End mill cutter Ø16–65	16	150	65	2	100

holders of various lengths and cutters with diameters Ø10 and Ø16 (Table 1). Unbalance measurements of the heat shrink sleeves used, with and without the tool attached, were also performed to determine the effect of tool overhang on its imbalance.

The manufacturer specifies all tool holders used as pre–balanced according to the balancing accuracy class G2.5, up to 25000 rpm. For tool Ø10 mm, the holders from Table 2, were used.

Table 2. Heat shrink holders for a tool with a diameter Ø10.

Nr	Holder name	Overhang [mm]
1	Ø10 × 80	80
2	Ø10 × 120	120
3	A6 Ø25 × 100 extension B3 Ø25 × Ø10 × 160	240

The tool holders for the Ø16 mm tool are described in Table 3.

Table 3. Heat shrink holders for a tool with a diameter Ø16.

Nr	Holder name	Overhang [mm]
1	A4 Ø16 × 80	80
2	A4 Ø16 × 200	200

The measurements were carried out in accordance with the procedure described by the machine manufacturer according to the corrective screws method due to the fact that each of the holders has for this purpose pre–made threaded holes. For each of the holders, three measurements were taken. As an efficiency criterion G2.5 allowable unbalance class was adopted, according to PN–93/N–01359, with a maximum tooling speed of 12000 rpm, selected on the basis of the maximum spindle speed of the available machine tool. The balancing speed was set as the manufacturer's recommended 1100 rpm. Firstly, an analysis was performed for the holders themselves, then for the holders with the cutting tool attached. Machining was carried out using the HAAS milling centre VF 3SS, on samples of 7075–T6 aluminium alloy, with wall thickness 20 mm, height and width 50 mm. Heat shrink holders with different overhangs were used. Milling tests were carried out for cases: holder without balancing process, for a balanced holder and

Table 4. Cutting parameters.

Nr	Tool name	V_c [m/min]	n [rpm]	f_z [mm/tooth]	a_e [mm]	a_p [mm]
1	Shoulder cutter Ø10–22	375	11 943	0.039	20	0.05
2	Shoulder cutter Ø16–65	350	6 967	0.058	24	0.05

for two with introduced imbalances. The cutting parameters for each tool were selected from the manufacturer's tool catalogues (Table 4).

The sample was clamped in a vice at a depth of $l = 20$ mm, machining was performed concurrently (Fig. 1). After clamping and before actual machining, the sample surface was pre–treated to ensure a constant radial infeed value for proper machining and to eliminate sample clamping error. One sample (number 0.0) was only pre–treated to obtain a reference surface for surface quality measurements.

a)   b)

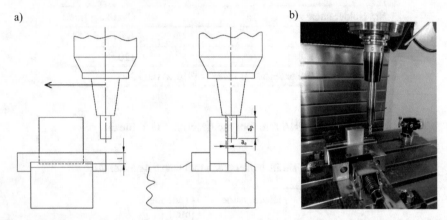

Fig. 1. Study on the impact of tool unbalances on surface quality. Process diagram (a) and the test photo (b).

To determine the unbalance, before the milling test each holder was unblocked from the machine tool to undergo an unbalance analysis. Prior to analysis, compensatory screws weighing 0.5 g were screwed into each of the holes of the holders for subsequent correction of the mass distribution. The milling procedure consisted of processing the samples in an unbalanced holder, then the holder was balanced by the method of compensating screws, by screws of varying weight and by changing their position so that the residual unbalance is as low as possible, after which a milling test with a balanced holder was carried out. The next step was to add unbalance by screwing a heavyweight compensation screw into one of the holes and perform a milling test with an unbalanced tool holder. The last procedure was to increase the unbalanced mass by increasing the weight of the imbalance screw and performing the last milling test. All steps were successively repeated for every holder. The measurement of surface roughness parameters was carried out using a Taylor Hobson Suronic 25 profilometer, over a 25 mm section, over a 100 mm range, using a 0.8 mm Gauss filter. The measurement was carried out three times for each sample at 2/3 of the axial infeed distance of the tool, in accordance with its feed direction.

3   Results

The results of unbalance measurements without the tool indicate an increase in unbalancing with an increase in tool overhang, with the exception of the Gühring Ø10 × 80 holder which alone does not meet this relationship and the requirements of the G2.5 standard. In the case of Ø10 tool holders, it can be seen that the tool with the largest overhang, despite the smallest unbalanced weight, has the greatest imbalance. Also in the case of the Ø16 tool holders, the tool with the largest overhang has a greater imbalance. This holder also, despite having smaller overhang than the longest holder Ø10, has the largest imbalance of all holders (Table 5).

Table 5. Unbalance measurement results of heat shrink holders without tool

Nr	Holder name	Tool diameter [mm]	Overhang [mm]	Unbalance [gmm]	Unbalanced mass [g]	Acceptable unbalance [gmm]
1	Ø10 × 80	10	80	2.5	0.15	2.1
2	Ø10 × 120	10	120	2.2	0.14	2.6
3	A6 Ø25 × 100 extension B3 Ø25 × Ø10 × 160	10	240	2.7	0.11	4.4
4	A4 Ø16 × 80	16	80	0.9	0.05	2.2
5	A4 Ø16 × 200	16	200	2.9	0.18	3.8

After mounting the tool, the unbalance did not increase only for the tool holder A4 Ø16 × 80, for the other holders there is a clear increase of unbalance which is due to the fact that the tool can introduce additional unbalance or change its value by increasing the overhang. The residual unbalance criterion G2.5 is only met by tool holder Ø16. As with measurements without a tool attached, you can see an increase in unbalance as the overhang increases. The largest imbalance is characterized by A6 Ø25 × 100 holder

Table 6. The results of unbalance measurements of heat shrink holders with the tool.

Nr	Holder name	Tool diameter [mm]	Overhang [mm]	Unbalance [gmm]	Unbalanced mass [g]	Acceptable residual umbalance [gmm]
1	Ø10 × 80	10	115	3.8	0.24	2.3
2	Ø10 × 120	10	155	3.9	0.25	2.8
3	A6 Ø25 × 100 extension B3 Ø25 × Ø10 × 160	10	275	4.9	0.21	4.5
4	A4 Ø16 × 80	16	180	0.9	0.05	2.9
5	A4 Ø16 × 200	16	300	3.3	0.20	4.5

with extension B3 Ø25 × Ø10 × 160, which also has the largest increase in unbalancing compared to the measurement result without the tool. The tool with the largest diameter, despite the greater length and weight, resulted in a small increase in unbalance (Table 6).

The last stage was the measurement of surface roughness parameters of milled samples with different imbalances. The following is a summary of selected roughness parameters (Table 7).

Table 7. Measurement results of geometrical structure parameters.

Tool	Holder	Unbalance [gmm]	Unbalanced mass [g]	Nr	Rp [μm]	Rv	Rt	Ra	Rq	Rz	Rsm	Wt
				0.0	0.490	0.407	1.575	0.170	0.207	0.897	0.145	2.448
End milling cutter Ø10–22	Ø10 × 80	4.4	0.35	1.1	0.388	0.353	1.028	0.140	0.170	0.741	0.154	2.308
		0.2	0.01	1.2	0.445	0.371	1.502	0.150	0.186	0.816	0.155	0.949
		19.9	1.63	1.3	0.395	0.341	1.204	0.144	0.174	0.736	0.163	1.649
		28.5	2.29	1.4	0.429	0.362	1.282	0.142	0.175	0.791	0.146	1.305
	Ø10 × 120	5.3	0.43	2.1	0.354	0.294	1.297	0.119	0.146	0.648	0.166	1.278
		0.1	0.01	2.2	0.402	0.327	0.986	0.132	0.162	0.729	0.141	0.985
		18.9	1.51	2.3	0.414	0.396	1.157	0.151	0.184	0.811	0.154	1.534
		35.9	2.92	2.4	0.412	0.354	1.100	0.136	0.166	0.765	0.140	1.282
	A6 Ø25 × 100	6.9	0.31	3.1	0.513	0.452	1.508	0.188	0.230	0.966	0.183	1.662
		0.1	0.01	3.2	0.504	0.439	1.581	0.183	0.223	0.943	0.179	1.136
		26.7	1.35	3.3	0.436	0.440	1.250	0.166	0.204	0.876	0.168	1.251
		42.2	2.19	3.4	0.455	0.408	1.215	0.165	0.202	0.863	0.165	1.131
End milling cutter Ø16–65	A4 Ø16 × 80	5.1	0.39	4.1	0.391	0.371	1.464	0.147	0.179	0.762	0.232	1.853
		0.0	0.00	4.2	0.368	0.322	1.193	0.129	0.159	0.689	0.186	1.307
		19.5	1.5	4.3	0.424	0.395	1.198	0.152	0.186	0.819	0.234	1.708
		42.5	3.27	4.4	0.413	0.355	1.196	0.148	0.182	0.769	0.208	1.822
	A4 Ø16 × 200	2.2	0.17	5.1	0.578	0.460	1.766	0.204	0.249	1.038	0.217	1.537
		0.2	0.02	5.2	0.585	0.507	1.615	0.221	0.264	1.092	0.244	1.537
		15.9	1.23	5.3	0.696	0.653	2.040	0.288	0.345	1.349	0.288	2.293
		46.0	3.55	5.4	0.709	0.607	2.010	0.277	0.332	1.316	0.263	2.627

There are slight differences in roughness parameters for each surface Rv, Rp, Rz not exceeding 0.1 μm and 0,05 μm Ra, Rq, Rsm, for surfaces machined with Ø10 holders, with the same overhang, regardless of balance. Comparison of Ra and Rz parameters for different tools and unbalances is shown in Fig. 2 and 3. For the A4 Ø16 × 200 holder, it can be seen that the imbalance introduced artificially increases the surface roughness parameters a little and between the machined surface with balanced and unbalanced holder no significant differences in roughness indicators can be seen. The parameter report also shows that holders with a longer overhang are more sensitive to imbalance, which is particularly evident in the A4 Ø16 × 200 holders, where the increase in surface roughness parameters of the machined surface after the introduction of imbalance was the largest. Both among Ø10 and Ø16 holders, the ones with the highest overhang have higher roughness parameters of the machined surface than the ones with the lowest overhang. It can also be seen that parameters Ra, Rq for A4 Ø16 × 200 holders are larger than for A4 Ø16 × 80 regardless of imbalance (similarly for the longest and shortest Ø10 holder).

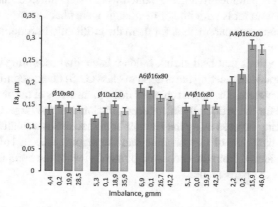

Fig. 2. The effect of tool unbalances on the roughness parameter Ra.

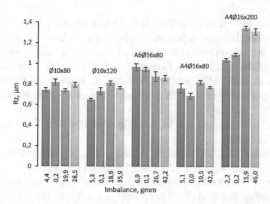

Fig. 3. The effect of tool unbalances on the roughness parameter Rz.

The lowest value of surface parameters Rv, Rp, Ra, Rq was obtained by a balanced A4 Ø16 × 80 holder, it also has the lowest unbalance value after the balancing operation. The highest values of roughness and corrugation parameters were recorded for the holder with the largest overhang A4 Ø16 × 200 with the largest, artificially introduced imbalance. The height of the wave profile has decreased (or has not changed with A4 Ø16 × 200) for a balanced holder relative to an unbalanced holder.

4 Conclusion

As a result of the tests, the effect of tool unbalance on surface quality in Al 7075–T6 alloy machining was determined.

- measurements of surface parameters machined with a tool with a different overhang and degree of unbalance showed that the unbalance values of the heat shrink holders used had little effect on the surface roughness parameters, for small overhangs the differences in the parameter values obtained were independent of the unbalance,
- surfaces made with tools with higher overhangs were characterized by higher values of roughness and wave parameters, for them the artificially introduced unbalance was also more important,
- the unbalance values that heat shrink holders have after mounting the tool, without balancing (despite not meeting the criteria of class G2.5) do not significantly affect the roughness parameters. The largest changes in the parameters of the geometric structure were recorded for the artificially introduced unbalance at the largest overhang.
- the tool balancing process using the Haimer Tool machine is a relatively simple operation, but the purchase of the machine and the significance of balancing the tools should be considered from the economic point of view of the company as well as the quality level of the parts made.

References

1. Sahare, S.B., Untawale, S.P., Chaudhari, S.S., Shrivastav, R.L., Kamble, P.D.: Experimental investigation of end milling operation on Al2024. Mater. Today: Proc. **4**(2), 1357–1365 (2017)
2. Young–Ho, N.: A study on surface roughness of Aluminium 7075 to nose radius and cooling method in CNC lathe machining. J. Korean Soc. Manuf. Process Eng. **14**(4), 85–91 (2015)
3. Borojević, S., Lukić, D., Milošević, M., Vukman, J., Kramar, D.: Optimization of process parameters for machining of Al 7075 thin-walled structures. Adv. Prod. Eng. Manag. **13**(2), 125–135 (2018)
4. Dursun, T., Soutis, C.: Recent developments in advanced aircraft aluminium alloys. Mater. Des. **56**, 862–871 (2014)
5. Baumann, J., Siebrecht, T., Wiederkehr, P., Biermann, D.: The effect of runout errors on process forces and tool wear. Procedia CIRP **79**, 39–44 (2019)
6. Cifuentes, E.D., Perez, G.H., Guzman, V.M., Vizan, I.A.: Dynamic analysis of runout correction in milling. Int. J. Mach. Tools Manuf. **50**, 709–717 (2010)
7. Wojciechowski, S., Twardowski, P., Pelic, M., Maruda, R.W., Barrans, S., Krolczyk, G.M.: Precision surface characterization for finish cylindrical milling with dynamic tool displacements model. Precis. Eng. **46**, 158–165 (2016)
8. ISO 16084:2017–Balancing of rotating tools and tool systems
9. ISO 1940–1:2003 Mechanical vibration – balance quality requirements for rotors in a constant (rigid) state – part 1: specification and verification of balance tolerances
10. PN–93/N–01359 – Mechanical vibrations – balancing of rigid rotors – determination of the permissible residual unbalance
11. López de Lacalle, L., Lamikiz, A., Fernández de Larrinoa, J., Azkona, I.: Advanced cutting tools. In: Davim, J. (eds.) Machining of Hard Materials. Springer, London (2011)
12. Machining and Monitoring Strategies. In: Cutting Tool Technology. Springer, London (2008)
13. Monies, F., Danis, I., Lagarrigue, P., Gilles, P., Walter, R.: Balancing of the transversal cutting force for pocket milling cutters: application for roughing a magnesium–rare earth alloy. Int. J. Adv. Manuf. Technol. **89**, 45–64 (2017)
14. Fleischer, J., Denkena, B., Winfough, B., Mori, M.: Workpiece and tool handling in metal cutting machines. Ann. CIRP **55**(2), 817–840 (2006). in German

15. Salgueroa, J., Batistaa, M., Calamazb, M., Girotb, F., Marcos, M.: Cutting forces parametric model for the dry high speed contour milling of aerospace aluminium alloys. Procedia Eng. **63**, 735–742 (2013)
16. Campatelli, G., Scippa, A.: Prediction of milling cutting force coefficients for Aluminum 6082–T4. Procedia CIRP **1**, 563–568 (2012)
17. Weinert, K., Surmann, T., Enk, D., Webber, O.: The effect of runout on the milling tool vibration and surface quality. Prod. Eng. Res. Dev. **1**, 265–270 (2007)
18. Franco, P., Estrems, M., Faura, F.: Influence of radial and axial runouts on surface roughness in face milling with round insert cutting tools. Int. J. Mach. Tools Manuf. **44**, 1555–1565 (2004)
19. Yue, C., Gao, H., Liu, X., Liang, S.Y., Wang, L.: A review of chatter vibration research in milling. Chin. J. Aeronaut. **32**(2), 215–242 (2019)
20. Żabiński, T., Mączka, T., Kluska, J., Kusy, M., Gierlak, P., Hanus, R., Prucnal, S., Sęp, J.: CNC milling tool head imbalance prediction using computational intelligence methods. In: Artificial Intelligence and Soft Computing. ICAISC 2015. LNCS, vol. 9119. Springer, Cham (2015)
21. Aguiar, M.M., Diniz, A.E., Pederiva, R.: Correlating surface roughness, tool wear and tool vibration in the milling process of hardened steel using long slender tools. Int. J. Mach. Tool Manuf. **68**, 1–10 (2013)
22. Alieva, R., Guseinov, R.: Self--balancing high–speed mill. Russ. Eng. Res. **37**(9), 784–788 (2017)
23. Haimer. Tool dynamic. Balancing for tool holders, grinding wheels and rotors. Operating manual

The Experimental Method of Determining the Forces Operating During the Abrasive Waterjet Cutting Process – A Mathematical Model of the Jet Deviation Angle

Tomasz Wala[✉] and Krzysztof Lis[✉]

Silesian University of Technology, Gliwice, Poland
{tomasz.wala,krzysztof.lis}@polsl.pl

Abstract. The article presents two mathematical models of the abrasive water jet deviation angle based on the interaction forces affecting the workpiece and the cutter head during cutting. One of the models is based on the reaction force component which remains in line with the direction of the feed. The other is based on the vertical reaction component (in line with the axis of the abrasive nozzle). In order to verify the models, experimental studies were carried out. Reactions of forces which affect the object were measured with the use of a four-component piezoelectric Kistler force sensor model 9272. The model based on the reaction force component which remained in line with the direction of the feed was positively verified. Tests have shown that real time measurements can be used to monitor the jet deviation, which is reflected in the topography of the cutting surface.

1 Introduction

Studies on surface topography after abrasive waterjet machining are carried out by many researchers. Understanding the nature of the formation of the surface topography permits developing a system to control the cutting process. Such a system should enable setting a range of process parameters in order to achieve/obtain desired surface quality in terms of its roughness and post-machining marks (striations, deformations). In their article [1] Valicek, Harnicarova and Kusnerova, observed that the wavy surface is a direct result of the abrasive water jet action on the workpiece and thus it carries information about the process parameters operating in the interaction between the machining tool and the workpiece. The research [2] by Zhao and Guo showed that the surface deformation process is largely influenced by the hardness of the workpiece. They also noticed that the smoothness of the cutting surface is demonstrably easier to obtain with harder materials. The process of material disintegration involves cutting mechanisms, plastic deformation, material fatigue and breakage. The article [3] presents research on surface topography, where the material removal mechanism was determined on the basis of tests. Ochsner and Altenbach were able to optimize the technological parameters of the abrasive water cutting process by using a function developed specifically to describe

© Springer Nature Switzerland AG 2020
G. M. Królczyk et al. (Eds.): IMM 2019, LNME, pp. 236–245, 2020.
https://doi.org/10.1007/978-3-030-49910-5_21

the surface topography. Through the use of three-dimensional surface methods (optical profilometer and optical microscope), surface analysis methods significantly expand the range of possibilities of assessing the surface texture, providing new information about the material removal mechanism. This was also confirmed by research carried out by Klichova, Klich and Zlamal, which they described in the article [4].

Also, Chen and Siores studied the cutting surface characteristics, which they described in the article [5]. By using a scanning electron microscope, the authors were able to show the influence of the abrasive particles distribution on the cutting surface (machined surface) deformation. This enabled the development of a technique to minimize the extent of deformation (waving).

As a consequence, the current state of knowledge allows for the development of alternative methods of determining the cutting conditions. These include methods for measuring the interactions occurring during the cutting process measured directly on the cutting head. Gou and Luis conducted extensive research in this area, the results of which were presented in article [6] and so did Chao, Chung and Geskin presenting their work in articles [9, 10]. Extensive research [11] on the impact of vibrations, their frequencies and the operating forces – measured at the cutter head – on the cutting surface quality carried out by Hreha Radvanska and Hloch create the possibility of correcting these conditions in real time. The measurement of acoustic emissions accompanying the process of waterjet cutting and the impact of changes in the cutting parameters on the cutting surface were also described by Momber, Mohan and Kovacevic in their article [12]. With reference to such methods, the authors of this article carried out their own exploratory tests in an attempt to identify the effects of the abrasive water jet action on the cutter head and the workpiece, which were described in studies [13, 14] and [15].

The research presented below focuses on real-time force action measurements and the use of a mathematical model describing how selected cutting force components affect the angle of the cutting jet. The angle of the cutting jet is directly reflected in the inclination of the post-machining marks on the surface after cutting, i.e. the striation angle (striation deviation).

2 Model of Jet Angle

The deviation of the jet during the process of cutting an object with abrasive waterjet partly depends on two velocities: the speed of the cutting jet coming out of the cutting head nozzle v_1 and the feed rate of the head during cutting in a specific direction u (Fig. 1).

By determining the components of velocity v_1 along the X and Y axes before the occurrence of the jet deviation phenomenon can be represented respectively as follows:

$$v_{1x} = u \tag{1}$$

$$v_{1y} = -v \tag{2}$$

Taking into account the deviation of the cutting jet, the jet velocity v_2 can be represented as components for the X and Y directions respectively as follows:

$$v_{2x} = (v + u) \cdot \sin \alpha \tag{3}$$

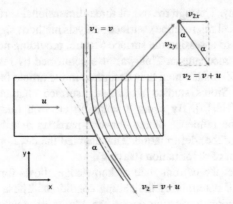

Fig. 1. The distribution of stream velocity vector components.

$$v_{2y} = -(v + u) \cdot \cos \alpha \tag{4}$$

Considering the forces (reactions) operating during cutting by means of a cutting jet, the dynamic reaction of the jet R (Fig. 2) can be determined from the principle of conservation of momentum of the jet, which is described by the vector:

$$\overline{R} = \rho Q(\overline{v}_1 - \overline{v}_2) \tag{5}$$

The components of the dynamic reaction vector in the coordinate system (X, Y) are respectively:

$$R_x = -\rho Q(\overline{v}_{1x} - \overline{v}_{2x}) \tag{6}$$

$$R_y = -\rho Q(\overline{v}_{1y} - \overline{v}_{2y}) \tag{7}$$

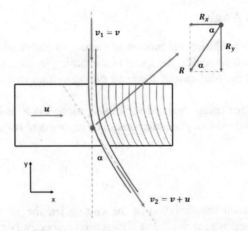

Fig. 2. Distribution of stream reaction components in the XY coordinate system during the cutting process.

Introducing the jet speed components into the coordinate system (X, Y), these reactions are respectively:

$$R_x = \rho Q((v + u) \cdot \sin \alpha - u) \tag{8}$$

$$R_y = \rho Q(v - (v + u) \cdot \cos \alpha) \tag{9}$$

The reaction component R_x is a force which occurs during the feed motion, which is accompanied by force F_f acting on the nozzle. The reaction component R_y is a force occurring along the axis of the cutting jet coming out directly from the cutting head (designated as the component F_h). By substituting the force values respectively and transforming the equations to determine the angle value for force F_f and F_h respectively, it can be represented that the value of angle α resulting from the operation of force F_f is:

$$\alpha = \arcsin \frac{\rho Q u + F_f}{\rho Q(v + u)} \tag{10}$$

and the value of angle α resulting from force F_h is:

$$\alpha = \arccos \frac{\rho Q u - F_h}{\rho Q(v + u)} \tag{11}$$

3 Experimental Determination of Force Reaction During the Cutting

A dedicated equipment (Fig. 3) consisting of an aluminum alloy base plate attached to the machine table (grating), on which a four-component Kistler force sensor model 9272 designed for measuring forces operating during machining was developed and made for the purposes of the research.

Fig. 3. Diagram and view of the test stand

An arm was attached to the mounting holes of the sensor. Sample workpieces were fixed at the end of the arm. The measuring track also contained a Kistler 5020 A piezo-electric amplifier dedicated to the sensor and a PXI class computer equipped with the NI PXI 6230 measuring card. The scheme of the measuring track is shown in Fig. 4.

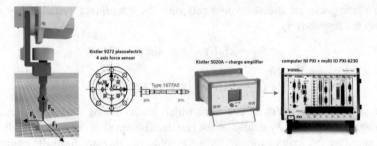

Fig. 4. The scheme of the measuring track

The sensor device was attached to the machine so that the direction of the sensor's X axis coincided with the direction of the machine, which corresponded to the direction of the feed force component F_f (Fig. 5). The sensor's Y axis direction coincided with the machine's Y axis direction and corresponded to the direction perpendicular to the feed direction F_b. Theoretically, the value of this force should be close to zero when cutting in a straight line, and its increasing values may result from the turbulence of the jet during the cutting process. The force measured along the Z axis corresponds to the axial force, i.e. F_h.

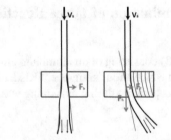

Fig. 5. Feed force component distribution F_f, F_h, F_b

During the experimental study, measurements were carried out for two types of materials. The selected materials used during the tests were aluminum alloy EN-AW-7075 and S235 RJ steel. The samples were 12 mm and 20 mm thick. The following machining parameters were used during the tests: pump pressure 330 MPa, abrasive flow rate 340 g/min. The head was fitted with a water nozzle with a hole diameter of 0.35 mm and an abrasive nozzle with a hole diameter of 1 mm.

The average values of forces and torques of the main part of the process, i.e. cuts in the entire cross-section of the material, were determined on the basis of graphs registered during force measurements, and selected results were summarized in Tables 1 and 2.

Table 1. Values of average reaction force during the cutting aluminum alloy EN-AW-7075 - 20 mm thick.

Feed rate [mm/min]	Force F_b [N]	Force F_f [N]	Force F_h [N]
98	0.03	3.99	2.35
137	0.57	3.93	2.34
217	0.46	9.06	3.31
304	0.12	10.57	9.46

Table 2. Values of average reaction force during the cutting S235 RJ steel - 20 mm thick.

Feed rate [mm/min]	Force F_b [N]	Force F_f [N]	Force F_h [N]
24	0.13	3.88	1.3
31	0.08	5.46	1.74
43	0.22	7.38	3.65
68	0.01	8.2	4.74
96	0.44	8.99	7.62

The graphs in Figs. 6, 7 show how the forces depended on the feed rate for individual samples.

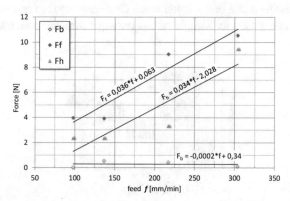

Fig. 6. Test results of the influence of cutting parameters on the forces acting on the nozzle. Cutting EN-AW-7075 aluminum alloy - 20 mm thick

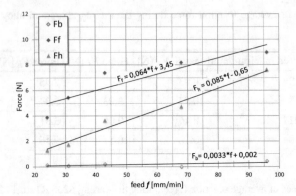

Fig. 7. Test results of the influence of cutting parameters on the forces acting on the nozzle. Cutting S235 RJ steel - 20 mm thick.

4   Result of Experimental Determination of Stream Deflection Angle

As part of the experimental research, measurements of the forces acting on the sample were made during the cutting of selected materials. Then, surface tests were carried out consisting in measuring the angle of deviation of post-machining marks, i.e. the deviation of the striations on the surface characteristic for abrasive waterjet machining (the angle between the axis of the nozzle and the tangent to the striations at the point of the jet at the exit point). The following fixed cutting parameters were used during the tests: pump pressure 330 MPa, abrasive flow rate) $m_a = 340$ g/min. The variable parameter was the feed rate of the cutter head. Changing the feed rate was necessary to force changes in the jet angle. Sample results for angle measurement tests are shown in Fig. 8.

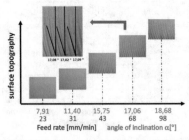

Fig. 8. Changing the topography of the cutting surface depending on the feed rate and the influence of the feed rate on the inclination angle - material S235RJ

Tables 3 and 4 contain examples of average force values measured by the piezoelectric sensor for one direction, i.e. the direction of the feed rate vector F_f. The tables summarize the results of the measured angle of deviation of post-machining marks in relation to the angle calculated according to Eq. (10) and to the measured average feed force F_f for various values of feed rate and the material cut, i.e. S235 RJ and EN-AW-7075.

Table 3. The comparison of the results of the measured angle value with the calculated angle value (S235 RJ)

Feed rate [mm/min]	Force F_f [N]	Measured angle [°]	Calculated angle [°]
24	3.88	7.91	7.961
31	5.46	11.40	11.24
43	7.38	15.75	15.27
68	8.2	17.06	17.02
96	8.99	18.68	18.72

Table 4. The comparison of the results of the measured angle value with the calculated angle value (EN-AW-7075)

Feed rate [mm/min]	Force F_f [N]	Measured angle [°]	Calculated angle [°]
98	3.99	8.26	8.18
137	3.93	8.20	8.06
217	9.06	18.74	18.87
304	10.57	22.43	22.16

Figures 9 and 10 show a comparison of the results obtained with the experimental method and the results obtained on the basis of the mathematical model of the angle of inclination of post-machining marks (striations) according to Eq. (10).

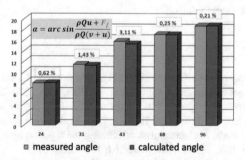

Fig. 9. The comparison of results of measured inclination angle with the results obtained on the basis of the model - material S235RJ

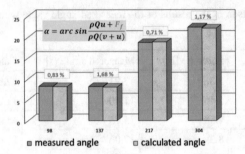

Fig. 10. The comparison of results of measured inclination angle with the results obtained on the basis of the model - material EN-AW-7075

5 Summary

The experimental studies presented in the paper allowed for the verification of two mathematical models presented in the further part of the paper. It has been shown that the model described by Eq. (10) with a significantly small relative error (up to ca. $\delta = 3\%$) reflects the actual values of the angle of inclination of the post-machining marks. The tests have shown that using the model described by Eq. (11), i.e. determining the angle based on the axial force F_h does not give satisfactory results. The relative error between the results predicted by the mathematical model described by Eq. (11) and the actual test results may even exceed $\delta = 100\%$. The high prediction rate offered by one of the proposed models of the post-machining striation inclination angle promises a possibility of monitoring the geometric parameters of the surface during the cutting process. However, this would require measuring the feed force F_f, e.g. by installing a sensor in the cutter head supporting structure. In the future, this should enable better control over the cutting process and permit automatic adaptation of the process parameters, making the cutting machines even "smarter":

References

1. Valicek, J., Harnicarova, M., Kusnerova, M., et al.: Propositon of a solution for the setting of the abrasive waterjet cutting technology. Meas. Sci. Rev. **13**(5), 279–285 (2013)
2. Zhao, W., Guo, C.W.: Topography and microstructure of the cutting surface machined with abrasive waterjet. Int. J. Adv. Manuf. Technol. **73**(5–8), 941–947 (2014)
3. Ochsner, A., Altenbach, H.: Mechanism of creating the topography of an abrasive water jet cut surface. Mach. Join. Modif. Adv. Mater. **61**, 111–120 (2016)
4. Klichova, D., Klich, J., Zlamal, T.: The use of areal parameters for the analysis of the Surface machined using the abrasive waterjet technology. In: Advanced in Manufacturing engineering and Materials, ICMEM 2018, Lectures Notes in Mechanical Engineering, pp. 26–44 (2019)
5. Chen, F.L., Siores, E.: The effect of cutting jet vibration on surface striation formation in abrasive water jet cutting. J. Mater. Process. Technol. **135**(1), 1–5 (2003)
6. Monno, M., Ravasio, C.: The effect of cutting head vibrations on the surfaces generated by waterjet cutting. Int. J. Mach. Tools Manuf **45**(3), 355–363 (2005)
7. Percel, V., Hreha, P., Hloch, S., Tozan, H., Valicek, J.: Vibration emission as a potential source of information for abarsive waterjet quality process control. Int. J. Adv. Manuf. Technol. **61**, 285–294 (2012)

8. Gou, N.S., Louis, H., Meier, G.: Surface structure and kerf geometry in abrasive waterjet cutting: formation and optimization. In: Seventh American Water Jet Conference Seatlle, Washington, 28–31 August, pp. 1–25 (1993)
9. Chao, J., Geskin, E.S., Chung, Y.: Investigation of the dynamics of the surface topography formation during abrasive waterjet machining. In: 11th International Symposium on Jet Cutting Technology St Andrews, Scotland, (8–10 September, 1992), pp. 593–603 (1992)
10. Chao, J., Geskin, E.S.: Experimental study of the striation formation and spectral analysis of the abrasive waterjet generated surfaces. In: 7th American Water Jet Conference Seattle, Washington, (28–31 August, 1993), pp. 27–41 (1993)
11. Hreha, P., Radvanska, A., Hloch, S., Perzel, V., Królczyk, G., Monkova, K.: Determination of vibration frequency depending on abrasive mass flow rate during abrasive water jet cutting. Int. J. Adv. Manuf. Technol. **77**, 763–774 (2015)
12. Momber, A.W., Mohan, R.S., Kovacavic, R.: On-line analysis of hydro-abrasive erosion of pre-cracked materials by acoustic emission. Theor. Appl. Fract. Mech. **31**(1), 1–17 (1999)
13. Wala, T., Lis, K.: Pomiary drgań podczas procesu cięcia strugą wodno-ścierną pod kątem oceny jakości powierzchni. Mechanik nr 8/9, s.1084–1085 (2016)
14. Wala, T., Lis, K.: Badanie oddziaływań siłowych podczas cięcia wysokociśnieniową strugą wodno-ścierną. Mechanik nr 8/9, s. 415–423 (2015)
15. Wala, T., Lis, K.: Identyfikacja częstotliwości i postaci drgań maszyny AWJ w aspekcie drgań głowicy. Mechanik nr 10, s. 876–878 (2017)

Application of Bayes Classifier to Assess the State of Unbalance Wheel

Krzysztof Prażnowski[✉] and Jarosław Mamala

Department of Vehicle, Opole University of Technology, Mikołajczyka5 Str.,
47-271 Opole, Poland
k.praznowski@po.opole.pl

Abstract. The article reports the results of a study concerned with determination of the state of unbalance of a pneumatic tire using an indirect method. For this purpose, the acceleration signal was recorded in the sprung mass. The time run of this signal was determined on the basis of the forces resulting from the effect of pneumatic tire unbalance that is further amplified by interference from the environment. One common type of interference from the environment includes road roughness, which affects the analyzed system in terms of varying the amplitude and frequency of the primary signal. In such a vibrating system, the key issue undertaken in the study is associated with determination of the effective way to be applied in the analysis of the data gained from measurement, and offer in objective assessment of the possibility of using the Bayes classifier to determine the state of tire unbalance in real road conditions. In the Bayesian approach to the analysis, the power and frequency of the analyzed signal provide important information that is largely relative to the length of the interval of the original signal applied in calculations. In the article, Paretto multi-criteria optimization process was employed to determine the optimal range of this time interval.

Keywords: Bayes classifier · Pneumatic tire · Sprung mass · Unbalance

1 Introduction

A vehicle is considered as a system of many masses connected with one another, among which we can distinguish sprung and un sprung masses that are coupled with each other by dissipative-elastic elements. The interactions of these masses result in changes in the value of vibration amplitude and a variations in the occurrence of the dominant frequency in the analyzed secondary measuring signal representing the acceleration of the sprung mass [11]. Additionally, the parameters of the contracting, elastic and damping materials vary in time, which additionally limits the possibility of their simple analysis due to the occurrence of random characteristics [1]. These changes result from wear and tear, temperature changes and aging resulting from the effect of the environment.

When we take on the analysis of the acceleration signal representing sprung masses in terms of its suitability for developing a criterion for inferring about a fault that is due to unbalance of a pneumatic tire, which forms an element of an unsprung mass, this signal

© Springer Nature Switzerland AG 2020
G. M. Królczyk et al. (Eds.): IMM 2019, LNME, pp. 246–258, 2020.
https://doi.org/10.1007/978-3-030-49910-5_22

is found to depend on the method of signal analysis. Two examples of the application would involve static analysis using probability theory [2] and spectral method (FFT, PSD). An important element in the process of the signal analysis is associated with the adoption of an appropriate time interval for the time run under analysis. This plays a particular role in the spectral method, as the frequency resolution of the Fourier transform is inversely proportional to the duration of the investigated time run. Hence, for the long time intervals applied in the analysis, a spectrum of a modulated signal is derived, and for short intervals, the total of spectra with various instantaneous frequencies of the harmonic components is obtained. Therefore, an important element of the analysis process needs to be concerned with objective determination of the time interval suitable for a given signal that will capture the significance of the analyzed signal by application of a mathematical method for its determination. This is not the only factor representing the complexity of the investigated problem, as the present authors experience gained from the identification of the frequency range of wheel unbalance in road conditions implies that such identification is particularly complex due to the resonance frequency in the vibrating system [2, 3, 7].

Therefore, as we know from the previous research and insights in the area [2, 5, 7, 8] concerned with the system analysis conducted for frequencies in the range up to $f_k < 10\,\text{Hz}$, the values of the amplitudes of the acceleration of sprung mass caused by a given unbalance equal to $m_{nw} = 0.08$ kg lead to the difficulties in determining whether a given force exerted on the system is feasible for the analysis as the signals form a common set of input data, and a direct assessment of the state of balance may be affected by a considerable classification error.

The force exerted by the pneumatic tire rolling on an uneven surface leads to its vibration and a dynamic load F_{zdyn} is exerted (Eq. 1). The level of both depends on the level of disturbance (degree of road roughness) as well as the technical condition of the pneumatic tire (ovalization error, unbalance, variable stiffness). Hence, the excitation force resulting from the unbalance (Eq. 2) causes the vibrations of the unsprung mass [2, 7].

$$F_{zdyn} = c(h - z) + cd \big/ \omega \big(\dot{h} - \dot{z}\big) \tag{1}$$

where:

    c - stiffness coefficient;
    h - level of road roughness
    z - wheel displacement
    d - coefficient representing run out of a tire
    w - angular velocity of the wheel.

$$F_{un} = m_{un} r_{un} \omega^2 \tag{2}$$

where:

    m_{un} - unbalanced mass
    r_{un} - radius of wheel unbalance

As a consequence, vibrations of the vehicle occur. Examples of recorded body vibration signals (sprung mass) for selected vehicle speeds, wheel balance and disturbances are provided in Figs. 1, 2.

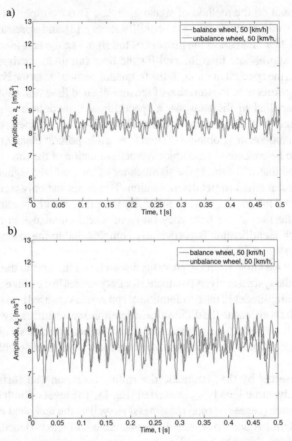

Fig. 1. Time runs representing vibrations of the chassis for the speed of V = 50 [km/h]: a) dynamometer test, b) road test.

Forces exerted by the rolling wheel are cyclic (harmonic). This is particularly evident in the case of dynamometer investigations that are free of road roughness. As the speed increases, the periods of individual signals become more dense. The analyzed unbalance is particularly visible for the speed of 90 km/h. In the time runs representing the amplitudes obtained for road tests, harmonic excitations are also visible, but their identification is difficult. The interference caused by rough roads makes it difficult to identify a fault in the wheel or the tire.

The research involving the analysis of the dominant frequency in the sample of the signal is associated with the need to use the numerical procedure of the Fourier transform. The Fourier transform is carried out for a specific number of samples n = 1, 2, 3, ..., N−1, where N is the length of the window calculated in the tests and resulting from the adopted time interval. Therefore, in the analysis we use a strictly defined time window with a finite number of observations (measurements), and its length, duration,

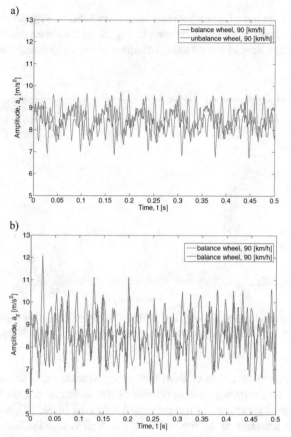

Fig. 2. Time runs representing vibrations of the chassis for the speed of V = 90 [km/h]: a) dynamometer tests, b) road tests.

time window is calculated as:

$$Tw = N/f_s, \tag{3}$$

where:

f_s - specific sampling frequency.

This operation involves the cut-off of a section of the analyzed signal without distorting it using a rectangular time window in quasi-stationary operating conditions of the tested system. This forms a very important stage from the point of view of determining the significance of a given signal. The phenomenon associated with the occurrence of components with frequencies that are close to the dominant frequency, which appear at the extreme sections of the time windows, hast be greatest effect on the occurrence of interference in the analyzed signal. Therefore, the use of a longer time window for the measurement provides a more accurate determination of the dominant frequency in the function of the signal spectrum. As a result, a more compact profile of the spectrum is derived, whereas the use of an excessively long time window suppresses the amplitude of the analyzed signal, which can be seen in Fig. 3. In the reverse case, the values of residual amplitudes around the dominant frequency are suppressed for too short time windows.

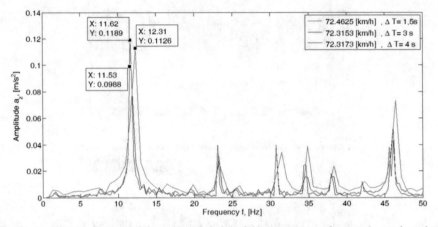

Fig. 3. Amplitude spectrum of the analyzed signal of the sprung mass for a study conducted on a dynamometer with a given unbalance with the unsprung mass $m_{un} = 0.08$ kg.

The use of excessively long or short measuring window leads to a change not only in the value of the frequency and amplitude of the dominant signal, but also affects the resulting signal spectrum. In the reports in this area [4, 9, 10], researchers agree that an incorrect selection of the time window can lead to significant differences in the assessment of the significance of a given signal. The lack of an unambiguous correlation between the variation in the length of the time window applied for the measurements and the value of the amplitude for the dominant frequency imposes the need to search for an objective method of selecting the measurement time window for the analysis. Therefore, in this paper multi-criteria optimization method (Pareto-Lorenz) was employed for the purposes of the analysis of the significance of the sprung mass acceleration signal with regard to the selection of the window parameters. The Pareto-Lorenz method comprises techniques that allows for an objective distinction for the most relevant factors affecting the analyzed set [12]. It provides a tool in the determination of optimal values for the adopted criteria, offering the increase of the significance of the analyzed signal, and in this case the criterion was adopted to be represented by:

maximum value of the amplitude of acceleration of the sprung mass,

value of the frequency of the force in the unsprung mass (f_k), in this case represented by the angular velocity of the wheel.

2   Pareto-Lorenz Method

The selection of the time window for measuring the amplitude of accelerations in the sprung mass of a passenger car was performed for constant windows of 1.5 s, 2 s, 2.5 s, 3 s, 3.5 s and 4 s. Examples containing the results for a passenger car traveling at a constant linear speed of 70 km/h (wheel angular speed $f_k = 11.28$ Hz, this is the speed affecting the forces acting on unsprung mass for the static radius of the pneumatic tire stat $= 0.275$ m) on a road with a bituminous surface, without visible damage, are presented in the form of a diagram (Fig. 4).

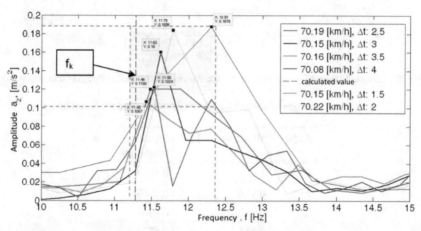

Fig. 4. Value of the acceleration amplitude of sprung mass in the function of the dominant frequency for the different durations of the time window

Due to the fact that the considered criteria are not mutually convertible and cannot be reduced to a single scalar criterion, rescaling was performed in relation to the minimum and maximum value of the considered criterion space. The scope of the criterion space for the data presented in Fig. 2 is equal to:

- for the frequency range from 11.43 Hz to 12.31 Hz, with a scalable value of 0–1.
- for the amplitude in the range from 0.1067 m/s^2 to 0.1876 m/s^2, with a scalable value of 0–1.

The resulting normalized vectors L were gained for the variable length of the measurement window Δt, as presented in Fig. 5.

The scalable value f(t) = 0 corresponds to the frequency of 11.43 Hz, whereas the value a(t) = 0 corresponds to the value 0,1876 m/s^2.

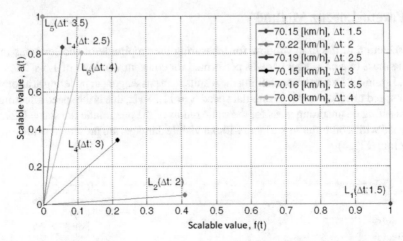

Fig. 5. Rescaled range of scalar criteria for the measurement window of Δt (scalable value)

Table 1. Length of vectors L_n for the adopted speed of 70 km/h.

Δt [s]	Vector length L_n	Signal magnitude
1.5	1	0.4835
2	0.412	0.4060
2.5	0.839	0.4761
3	**0.404**	**0.4725**
3.5	1	0.4761
4	0.814	0.2601

For the data developed there in order to objectively select the time window applied for measurements, the values of the signal power were determined for the dominant frequency over the selected lengths of the time window Δt. The results are presented in Table 1. The highest power signal value was obtained for $\Delta t = 1.5$ s (0.4835 m/s^2), whereas the lowest was recorded for delta $\Delta t = 4$ s (0.2601 m/s^2). The second criterion is associated with the minimum length of the scalar vector Lp, which for the time window $\Delta t = 1.5$ s is not taken into account despite the fact that highest signal power was obtained in these conditions. In this case, the time window $\Delta t = 3$ s has the best scalar vector Lp, which does not represent the mean value of the signal in terms of signal power and is 2.3% smaller from the maximum value. However, in this case, the signal power is 81.7% greater than the minimum value recorded for the time window $\Delta t = 4$ s. Based on the analysis of the length of the resulting vectors of the scalable criterion functions and the determined signal powers, the most favorable measurement window was determined, which for the analyzed linear vehicle speed of 70 km/h was found to be equal to 3 s.

We can also note at this point that similar results were obtained for the remaining linear speeds of the vehicle, however, in certain measuring conditions.

The same methodology was followed and tested in other conditions for which not only the linear speed of the vehicle was varied but the conditions of the experiment changed as well. The analysis of the results gained for the input data demonstrate that the length of the measurement window of the primary signal significantly affects the value of the signal as well as the harmonic dominant frequency that occurs in it. The results obtained for individual vehicle linear speed tests are presented in Tables 2, 3, 4, and 5.

Table 2. Length of vector L_n in the conditions of the test: dynamometer test, balanced wheel

Δt [s]	V = 50 [km/h]		V = 70 [km/h]		V = 90 [km/h]	
	L_n	Power signal $[m^2/s^4]$	L_n	Power signal $[m^2/s^4]$	L_n	Power signal $[m^2/s^4]$
1.5	1	0.0016	1.0833	0.0007	1.414	0.0004
2	0.842	0.0013	0.4659	0.0007	0.647	0.0012
2.5	0.964	0.0012	0.3047	0.001	0.361	0.0014
3	**0.773**	0.0012	1.0148	0.0004	0.202	0.0014
3.5	0.841	0.0012	**0**	0.001	**0.186**	0.0012
4	1	0.0011	0.5187	0.0007	0.444	0.0009

Table 3. Length of vector L_n in the conditions of the test: dynamometer test, unbalanced wheel with the mass of 0.08 kg

Δt [s]	V = 50 [km/h]		V = 70 [km/h]		V = 90 [km/h]	
	L_n	Power signal $[m^2/s^4]$	L_n	Power signal $[m^2/s^4]$	L_n	Power signal $[m^2/s^4]$
1.5	1.018	0.0004	1.414	0.011	1.048	0.0239
2	0.588	0.00045	0.688	0.0135	1.180	0.0342
2.5	**0.519**	0.00038	0.711	0.0139	0.628	0.0475
3	0.879	0.00031	0.829	0.0121	0.333	0.0557
3.5	1.003	0.00029	**0.246**	0.0176	**0.139**	0.0547
4	0.976	0.00029	0.285	0.0156	0.257	0.0462

Table 4. Length of vector L_n in the conditions of the road test: bituminous surface with good surface, balanced wheel.

Δt [s]	V = 50 [km/h]		V = 70 [km/h]		V = 90 [km/h]	
	L_n	Power signal [m^2/s^4]	L_n	Power signal [m^2/s^4]	L_n	Power signal [m^2/s^4]
1.5	1.414	0.0002	1.118	0.0062	1.04	0.0084
2	0.603	0.0004	1.2005	0.0077	0.571	0.0105
2.5	0.354	0.0004	0.5386	0.0085	**0.427**	0.0083
3	0.729	0.0002	0.5389	0.0105	0.521	0.007
3.5	0.392	0.0003	0.5522	0.0083	0.893	0.0048
4	**0.213**	0.0003	**0.2895**	0.0112	1.027	0.0042

Table 5. Length of vector L_n in the conditions of the road test: bituminous surface, no faults, unbalanced wheel with a mass of 0.08 kg

Δt [s]	V = 50 [km/h]		V = 70 [km/h]		V = 90 [km/h]	
	L_n	Power signal [m^2/s^4]	L_n	Power signal [m^2/s^4]	L_n	Power signal [m^2/s^4]
1.5	1	0.0032	0.6471	0.0729	1.414	0.0292
2	**0.292**	0.0025	1.4142	0.0467	0.799	0.0317
2.5	1.001	0.0011	0.3922	0.063	0.642	0.0317
3	1.137	0.0019	0.8571	0.0488	0.384	0.0328
3.5	0.756	0.0018	0.2973	0.061	0.289	0.0331
4	0.629	0.0018	**0.2114**	0.0593	**0**	0.0346

The results of the scalar vector analysis presented above demonstrate that the scalar vector is not strictly assigned to a given window representing measurement time, but it tends to vary depending on variables. For instance, it is relative to the speed of the vehicle and also depends on the conditions of the test. Accordingly, for the case of the tests on dynamometer free of road roughness, the value of the L_n scalar vector increases and then maintains a constant value (3.5 s). This value is independent of the state of unbalance imposed on the unsprung mass (Fig. 6a). As a result of changing the test conditions, in which road tests were conducted, the forces acting on the wheel vary with the linear speed of the vehicle also for the case of the balanced wheel. For a vehicle speed below 70 km/h, the scalar vector value changes $\Delta T = 2$ s to $\Delta T = 4$ s (Fig. 6b).

However, in this study it was found that the calculated values of the scalar vector Ln for the investigated linear speeds of the vehicle during the road test and the given unbalance in the unsprung mass did not maintain the regularities specified above. For the test vehicle linear speed of 90 km/h, the scalar vector value varied from $\Delta T = 4$ s to $\Delta T = 2.5$ s (Fig. 6b).

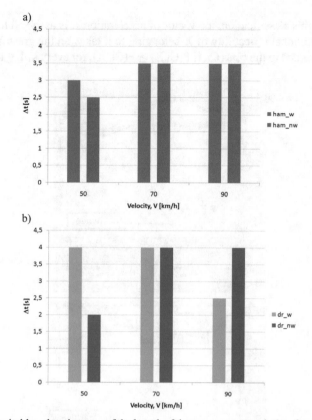

Fig. 6. Most suitable values in terms of the length of the measurement window for selected vehicle speeds

3 Determination of Signal Significance

In order to determine the conditions for the objective classification of the significance of the analyzed signal of the unsprung mass acceleration, resulting from the forces exerted by the unsprung mass with a given unbalance, and thus offering an objective assessment of the tire imbalance, a further analysis was carried out by application of the primary data recorded during the experiment. Signal classification analysis was performed in a two-dimensional observation space (in the frequency – amplitude domain) applying using a statistical classifier based on the Bayes theorem. This method is based on a direct statistical approach which is described by Eq. (2). It offers the prediction of the probability P(C|X) representing the fact of the analyzed signal with properties X belonging to a given class C.

$$P(C|X) = \frac{P(X|C)P(C)}{P(X)} \tag{4}$$

Each analyzed signal is considered as vector X with the values of attributes (properties) A1, ..., An, whereas C1, ..., Cn are subsets called classes to which X may belong.

As a result of this classification, the vector of acceleration X is assigned to the class to which the conditional probability of X belonging to it takes on the greatest value. The vector X is assigned to the class Ci, if $P(Ci|X) \geq P(Ck|X)$, for every k, $1 \leq k \leq m$, $k \neq i$.

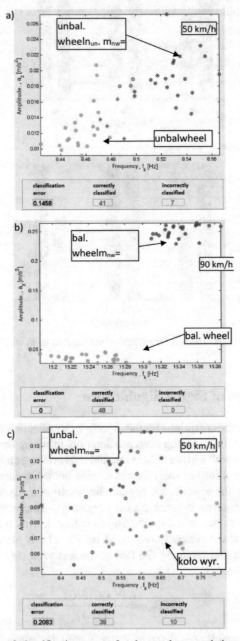

Fig. 7. Determination of classification errors for the results recorded on a dynamometer (a, b) and in the road test (c, d)

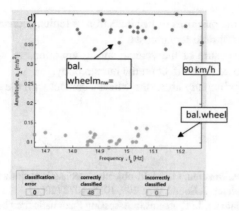

Fig. 7. (*continued*)

The application used for data classification was developed in the Matlab computational environment and the analysis was carried out for sets according to the same length of the measurement time window and was repeated over 24 trials, resulting in a priori probability of classification of 0.05. As a result of the classification, quantitative information on the data correctly and incorrectly classified into individual classes was obtained, as well as the value of the classification error. Examples of data classification results for a balanced tire with a specified unbalance of 0.08 kg are presented in Fig. 7. The details in Figs. 7a,b focus on the results of dynamometer tests and Figs. 7c,d presents the results for road tests, carried on a bituminous road surface, without visible roughness.

Figure 7 contains data regarding the extreme values of the linear vehicle speeds, i.e. 50 and 90 km/h, which correspond to the frequency of exerted force for the unsprung mass equal to $f_k = 8.5$ Hz, therefore, conducted when the condition $f_k < 10$ Hz is fulfilled. The analyzed data does not provide a direct guide applicable for the purposes of classification, although it can be seen clearly that the same conditions accompanies the collection of the data. The classifier employed in these conditions is characterized by mutual overlapping of classes, which results in the data classification error of 14.58% for the dynamometer tests and 20.83% for road tests. In this case, the Bayes classifier used only indicates the possibility of wheel unbalance and forms an indication to conduct further diagnostics of the vehicle's chassis. The use of Bayes classifier for the frequency satisfying the condition $f_k > 10$ Hz; therefore, for the frequency of excitation $f_k = 15$ Hz, the correct classification of all analyzed tests was performed both in the case of dynamometer and road tests. In this case, the classification error was 0% for 24 tests, despite the differences in the primary signal amplitude between the tests conducted on the dynamometer and on the road

4 Summary

The applicability of Bayesian classifier for determination of the state of unbalance of a pneumatic tire of a passenger car in real road conditions forms an up-to-date research issue for several reasons:

- economy, serving the purposes of fast detection of faults, decrease of operating costs of the vehicle and possible repair costs;
- comfort, longer exposure of the vehicle spring vibrations to which passengers are subjected to has an adverse effect on the human body;
- safety, as it can significantly affect the wheel's contact with the road surface.

References

1. Gobbi, M., Levi, F., Mastinu, G.: Multi-objective stochastic optimization of the suspension system of road vehicles. J. Sound Vibr. **298**, 1055–1072 (2006)
2. Prażnowski, K., Mamala, J.: Problems in Assessing Pneumatic tire Unbalance of a Passenger Car Determined with Test Road in Normal Conditions, SAE Technical Paper 2017-01-1805 (2017)
3. Mamala, J., Prażnowski, K., Augustynowicz, A.: Assessment of roadway surface condition based on measurements of sprung mass acceleration in the conditions of actual drive of a passenger car. In: 21th International Scientific Conference on Transport Means – 2017, Judokrante, Litwa (2017)
4. Bendat, J.S., Piersol, A.G.: Random Data. Analysis and Measurement Procedures. Wiley, New York (2010)
5. Byoung, S.K., Chang, H.C., Tae, K.L.: A study on radial directional natural frequency and damping ratio in a vehicle tire. Appl. Acoust. **68**, 538–556 (2007). Elsevier
6. Johnson, T.J., Adams, D.E.: Composite indices applied to vibration data in rolling tires to detect bead area damage. Mech. Syst. Sig. Process **21**(5), 2161–2184 (2007)
7. Pacejka, H.B.: Tire and Vehicle Dynamics. Butterworth-Heinemann, Oxford (2006)
8. Yunta, J., Garcia-Pozuelo, D., Diaz, V., Olatunbosun, O.: Influence of camber angle on tire tread behavior by an on-board strain-based system for intelligent tires. Measurement **145**, 631–663 (2019)
9. Chen, K.F., Cao, X., Li, Y.F.: Sine wave fitting to short records initialized with the frequency retrieved from Hanning windowed FFT spectrum. Measurement **42**(1), 127–135 (2009)
10. Andria, G., Savino, M., Trotta, A.: FFT-based algorithms oriented to measurements on multifrequency signals. Measurement **12**(1), 25–42 (1993)
11. He, S., Tang, T., Ye, M., Enyong, X., Deng, J., Tang, R.: A domain association hierarchical decomposition optimization method for cab vibration control of commercial vehicles. Measurement **138**, 497–513 (2019)
12. Beruvides, G., Castaño, F., Quiza, R., Haber, R.E.: Surface roughness modeling and optimization of tungsten–copper alloys in micro-milling processes. Measurement **86**, 246–252 (2016)

Theoretical and Actual Feed Per Tooth During Wood Sawing on an Optimizing Cross-Cut Saw

Kazimierz A. Orlowski[1(✉)], Wojciech Blacharski[1], Daniel Chuchala[1], and Przemyslaw Dudek[2]

[1] Department of Manufacturing Engineering and Automation, Faculty of Mechanical Engineering, Gdansk University of Technology, 11/12 Narutowicza Street, 80-233 Gdansk, Poland
kazimierz.orlowski@pg.edu.pl
[2] REMA S.A., 5 Boleslawa Chrobrego Street, 11-440 Reszel, Poland

Abstract. The work presents the geometries of circular saw teeth, whose task is to divide the cross-cut of the cutting layer. For efficient machining on a cross-cut saw with high feed speed values (about $1\ \mathrm{ms}^{-1}$) maximal theoretical feeds per tooth are determined. These values were compared with the actual values determined during transverse cutting of pine wood on a special test stands in industrial conditions. In the experiment the rotational speed of the main driving motor and the feed speed of the saw blade were simultaneously measured. Unique experimental results revealed that the feed driving system with the crank mechanism driven with the rotary servo motor only the phenomenon of the decrease in the spindle rotational speed was present, which caused a slight increase of the feed per tooth. In the case of the pneumatic actuator applied in the feeding driving system its flexibility together with larger changes in feed speed additionally were observed. For that reason in a new cross-cut saw for the feeding system driven by the crank mechanism with the rotary servo motor ought to be recommended.

Keywords: Optimizing cross-cut saw · Circular saw blade · Feed per tooth · Cutting

1 Introduction

Optimizing cross-cut saws, also called optimizers, belong to the group of cross-cut sawing machines often used in wood processing plants, and their purpose is to align the ends of lumber, trim to length and fast and precise cut out the anatomical defects during cross-cutting of single boards in solid wood. Thanks to the use of modern cross-cut saws and effective optimization for defect removal, it is possible to expect a reduction of raw material waste up to 8% [1].

Modern optimizing cross-cut saws are very dynamic machine tools, in which circular saw blades work with cutting speeds of about $80\ \mathrm{m\cdot s}^{-1}$, and average feed speeds can exceed $2\ \mathrm{m\cdot s}^{-1}$ (maximum up to $3.7\ \mathrm{m\cdot s}^{-1}$). The latter gives the cycle time of cutting approximately 0.085 s in the Opti Cut 450 Quantum model [1]. On the other

© Springer Nature Switzerland AG 2020
G. M. Królczyk et al. (Eds.): IMM 2019, LNME, pp. 259–267, 2020.
https://doi.org/10.1007/978-3-030-49910-5_23

hand, tool manufacturers report about cycle times in the range from 0.3 s to 1 s [2]. In the fastest sawing machines, the raw material movement system allows speeds of up to 415 m·min^{-1}, and exact positioning from full speed followed by re-acceleration is performed at up to 50 m·s^{-2}. This kind of machine cuts with an accuracy of up to ± 0.8 mm [1]. It should be emphasize that in printed materials as well as internet sources of many manufacturers, the speed of material movement is erroneously called the feed speed, which is in contradiction with the definition of this movement in the standard ISO 3002-3 [3].

Nasir and Cool [4] investigated the effect of feed speed, rotation speed, depth of cut, and the average chip thickness on the power consumption and waviness during the circular sawing process of green Douglas-fir wood. On the other hand, Mohammadpanah and Hutton [5] examined a stability analysis of spinning disk, an idealized representation of circular saw, for collared and splined arbor saws, when subjected to radial and tangential in-plane forces. Although the results described in works [4, 5] are valuable they cannot be utilized in the machine tools, which are the subject of this paper, since, in optimizing cross-cut saws fixed length cutting, defect cutting and optimization is conducted with extremely fast saw strokes, therefore, some saw blades producers call this method cutting "a punching method" [6]. The method for determining energetic effects for cross-cutting of wood is presented in the paper [7]. Furthermore, the method of selecting the desirable angular position of the cutting aggregate arm in the structural layout of the machine is also explained [7].

The Polish manufacturer of machine tools for woodworking Rema SA (Reszel, PL) offers an optimizing cross-cut saw with an automatic pusher (Fig. 1), in which a feed movement is realized by means of a pneumatic actuator. Nevertheless, latterly the company has been financially supported by the European Regional Development Fund, and the improvement of the feeding system of the machine is one of the project goals. Hence, in this paper the theoretical and actual values of the feed per tooth for two feed systems for the pneumatic and the crank mechanism driven with the rotary servo system are compared.

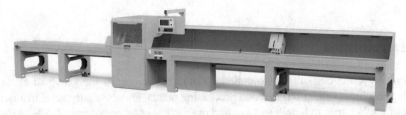

Fig. 1. The optimizing cross-cut saw with an automatic pusher CASTOR 500 currently produced by Rema SA

2   Theoretical Background

In the modern feed systems of optimizing cross-cut saws the pendulum systems are mainly applied, in which the spindle moves on a curved trajectory [8–10]. The feed motion can be realized by means of a pneumatic actuator (Fig. 2a) [11] or in newer solutions due to crank mechanisms driven with the rotary servo motor (Fig. 2b) [1, 12], which are called as ultra-rapid cutting units.

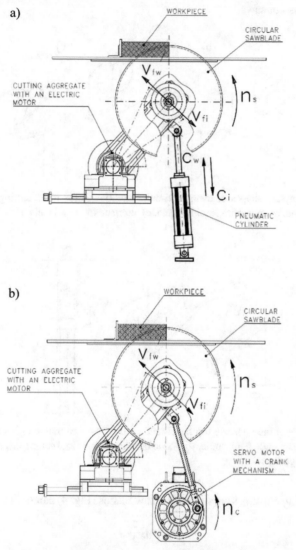

Fig. 2. The pendulum systems of optimizing cross-cut saws with the feed motion realized by means of a pneumatic actuator (a) and due to the crank mechanism driven with the rotary servo motor (b), where: v_{fw} – working feed speed, v_{fi} – idling feed speed, n_s – rotational speed of the circular saw blade, n_c – crank rotational speed, c_w – piston rod speed in the working stroke, c_i – piston rod speed in the idling stroke

Circular saw blades with sintered carbide inserts or stellite tipped teeth are usually used in woodworking machines. The exemplary teeth shapes are shown in Fig. 3. Figure 4 presents the geometry of alternate top bevel teeth (cutting tool geometry according to ISO standards [13]), which is commonly used in saw blades for efficient cutting on optimizing cross-cut saws.

a)   b)   c)

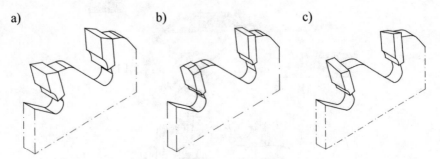

Fig. 3. The exemplary shapes of teeth of circular saw blades for wood cutting, where: square teeth regular shape, b) square/trapezoidal teeth, c) alternate top bevel teeth

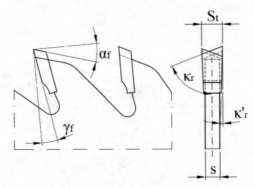

Fig. 4. Geometry of the alternate top bevel teeth, where: S_t – overall set (theoretical kerf), s – blade thickness, α_f – side flank angle, γ_f – side rake angle, κ_r – tool cutting edge angle, κ_r' – tool minor cutting edge angle

Feed speed v_f for this kind of circular saw blades (Fig. 4) can defined as follows:

$$v_f = f_z \frac{z}{2} n_s \tag{1}$$

where:

f_z – feed per tooth,

z – teeth number of the circular saw blade,

n_s – rotational speed of the circular saw blade.

After the transformation of Eq. (1) the feed per tooth is expressed as:

$$f_z = \frac{2v_f}{z \cdot n_s} \tag{2}$$

The theoretical maximal value of feed per tooth could be determined from Eq. (2) under assumption that the working feed speed $v_f = v_{fmax}$ and simultaneously rotational speed of the circular saw blade n_s is constant.

During sawing the value of feed per tooth might change as a result of:

- an effect of the feed system sensitiveness, which in case of the pneumatic actuator probably smaller values of feed speed could appear. Nevertheless, larger stiffness in case of the system with the crank mechanism driven with the rotary servo motor the mentioned effect is not expected;
- decrease in the spindle (circular saw blade) rotational speed caused by changes in feed speed. If the electric motor is controlled with the variable speed drive VSD (frequency converter) the behaviour of the driving system could be affected by the motor control mode either the V/f (Voltage/frequency) mode or the SLVC (Sensor Less Vector Control) mode. The latter should be applied if the application requires more dynamic and speed assurance.

3 Materials and Method

The tests of the cross-cutting process were carried out on two prototype stands under industrial conditions at REMA S.A. in Reszel (PL). The pendulum systems of optimizing cross-cut saws with the feed motion realized by means of a pneumatic actuator (Fig. 2a) and due to the crank mechanism driven with the rotary servo motor (Fig. 2b) were under investigations. On the both prototype stands rotational speeds of the electric engines were measured with the use of the incremental encoders type DFS60B-S4UA08192 (f. Sick (D)) (Fig. 5a). the same types of encoders were applied for measurement of the rotational speeds of the arms of the sawing aggregates (Fig. 5b). The encoder shafts were joined with the examined systems by bar couplings made of the glass fiber-reinforced polyamide (in the foreground in Fig. 5a).

a)

b)
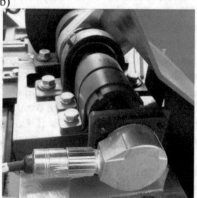

Fig. 5. Mounted incremental encoders for measurements of the both rotational speeds of the electric engine of the main drive (a) and the arm (in the background) of the sawing aggregate (b)

The spindles of the cutting aggregates are driven by the electric motors (nominal power $P_n = 7.5$ kW, nominal speed $n_n = 2830$ rpm, f. Tamel (PL)) by the toothed belt transmission with the ratio equal to $i = 1.26$. Therefore, the circular saw blade could rotate with the rotational speed equal to $n_s = n_n \cdot i = 3565.8$ rpm. The electric motor is controlled with the variable speed drive VSD type VLB3 7.5 kW (f. Lovato, Italy), and could be controlled either in the V/f (Voltage/frequency) mode or the SLVC (Sensor Less Vector Control) mode.

The crank mechanism was driven by the servo motor (three-phase synchronous motor) type 8LSA75DA030S000-3 (f. B&R, Austria) with technical data: nominal speed of the motor $n_n = 3000$ rpm, the nominal torque $M_n = 30$ Nm and maximum torque $M_{max} = 187$ Nm.

In experiments pine (*Pinus sylvestris* L.) samples of moisture content MC 15.7%, with dimensions in the cross section of 200 mm × 60 mm, were sawn. Technical data of the circular saw blade applied is as follows: outside diameter $D = \varnothing450$ mm, hole diameter $d = \varnothing30$ mm, blade thickness $s = 3.5$ mm, total overall set (theoretical kerf) $S_t = 4.8$ mm and teeth number $z = 138$. Sintered carbide teeth have had geometry shown in Fig. 4, with angle values: flank side angle $\alpha_f = 15°$, rake side angle $\gamma_f = 10°$ and tool cutting edge angle $\kappa_r = 70°$. The circular saw blade was clamped on the spindle by collars with the diameter equal to $\varnothing112$ mm. For comparison purposes, the maximum feed speed for two examined feed drive systems was expected to be 1 m·s^{-1}. Hence, the estimated maximal theoretical value of feed per tooth, for this cutting conditions, was equal to $f_{zmax} = 0.004$ mm.

In the experiments a software, which controlled cooperation of the USB measurement modules and incremental encoders, designed in the LabVIEW SignalExpress environment has been applied. Sampling rate in the experiment was equal to 2 kHz.

4 Results and Discussion

Figure 6 presents the spindle rotational speed and feed speed in a function of time while the feed motion was realized by means of a pneumatic actuator.

In Fig. 6a courses of the spindle rotational speed and the feed speed are shown. The plot of the feed speed (Fig. 6a) consists of two parts, the first presents changes of feed speed while the arm of the cutting aggregate (Fig. 2a) moves up with the working feed speed v_{fw}, and the second part when the is the idling stroke with the feed speed v_{fi}. On the other hand, Fig. 6b shows courses of the spindle rotational speed and the feed speed during cross cutting of the pine lumber of dimensions in the cross section 200 mm × 60 mm. In this experiment the main electric motor controlled with the variable speed drive VSD type VLB3 7.5 kW (f. Lovato, Italy) worked in the V/f (Voltage/frequency) mode. For that reason, even for a small value of the feed speed equaled during cutting $v_{fw} = 0.23$ m·s^{-1}, the spindle rotational speed rapidly decreased at the begging of sawing and at the time of leaving the workpiece the dynamical changes of the spindle rotational speed are observed.

Moreover, the comparison of feed speed changes revealed that the relative drop in the feed per tooth f_z caused by the flexibility of the pneumatic actuator was equal to about 9.6%, whereas the reverse phenomenon was observed in case of the spindle rotational

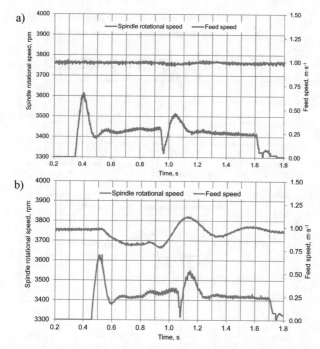

Fig. 6. Spindle rotational speed and feed speed while the feed motion was realized by means of a pneumatic actuator, where: a – idling, b – during cutting of pine lumber (200 mm × 60 mm)

speed decrease. For the latter, the drop in the spindle rotational speed caused increase of the feed per tooth equaled to 2.2%. The resultant change in the relative feed speed was −7.4%.

Figure 7 presents the spindle rotational speed and the feed speed in versus time while the feed motion was realized by means of the crank mechanism driven with the rotary servo motor. In this case the feed drive requires more dynamic and speed assurance, hence, the SLVC (Sensor less Vector Control) mode for the variable speed drive VSD was applied. In the working stroke some kind of deformation of the feed speed course at the time of the circular saw blade entry into the workpiece was observed (Fig. 7b), and simultaneously at this time the drop in the spindle rotational speed is noticed. This phenomenon caused an increase in feed per tooth equal to 6.36%. A comparison between maximal feed speeds during the idling (Fig. 7a) and working cycles (Fig. 7b) shown that the speeds are at the same level ($1.09\ \mathrm{m\cdot s^{-1}}$).

For an assessment of the quality of the feed drive system it is proposed the use of the specific change in feed speed, which might be defined as a ratio of the fixed nominal (idling) feed speed to the relative change of feed per tooth. The carried out assessment revealed that the specific change in feed speed for the feed system with the pneumatic actuator was $-0.035\cdot\mathrm{m\cdot(s\cdot\%)^{-1}}$, whereas in the case of the drive system with the crank mechanism was $+0.171\ \mathrm{m\cdot(s\cdot\%)^{-1}}$. The latter result is a proof that the feed drive with

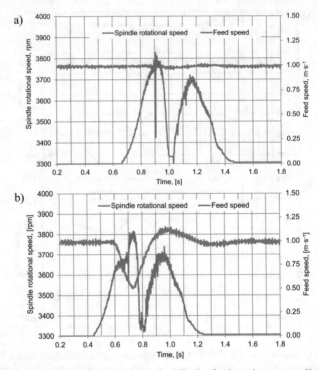

Fig. 7. Spindle rotational speed and feed speed while the feed motion was realized by means of the crank mechanism driven with the rotary servo motor, where: a – idling, b – during cutting of pine lumber (200 mm × 60 mm)

the crank mechanism driven with the rotary servo motor together with the SLVC mode applied in the spindle drive is a better solution in comparison to the feed drive with the pneumatic actuator.

5   Conclusions

The carried out analyses allowed us to discover undesired effects in the form of cutting conditions changes while sawing on the optimizing cross-cut saw.

- It was observed that the actual feed per tooth value could be affected by the flexibility of the both the feed drive system and the applied mode of the variable speed drive VSD. The both phenomena were simultaneously noticed during cutting while the cutting aggregate was driven by the means of a pneumatic actuator, and the resultant change in the relative feed speed was −7.4%.
- In the case of the feed drive system with the crank mechanism driven with the rotary servo motor only the phenomenon of the increase of the feed per tooth was observed. The latter was caused by a decrease in the spindle rotational speed. This phenomenon caused an increase in feed per tooth equal to 6.36%.

- Since, in the case of the feeding system driven by the means of a pneumatic actuator the set-up of the feed speed is difficult and its changes are unpredictable because of the system flexibility, hence, in the future in a new cross-cut saw the crank mechanism driven with the rotary servo motor should be applied. Furthermore, for the latter driving system feed speeds during the idling (Fig. 7a) and working cycles (Fig. 7b) have shown that the speeds are at the same level ($1.09 \ \text{m}\cdot\text{s}^{-1}$).

Acknowledgements. It is kindly acknowledged that this work has been carried out within the framework of the project POIR.01.02.00-00-0091/17, which has been financially supported by the European Regional Development Fund. The authors also would like acknowledge the company REMA S.A. in Reszel (Poland), which is the beneficiary of this project.

References

1. Dimter OptiCut 450 Quantum. https://falkenberg.no/media/produkter/produktark/sagbruk/73-prospekt-opticut-highspeed-serie-gbr.pdf. Accessed 8 May 2019
2. Sawing Leitz Lexicon (edn. 7). 01 Sawing website.pdf. http://www.leitz.org/?action=document&invisible=1&id=1800&PHPSESSID=kinvbesobbfs1qg07cduknso71. Accessed 8 May 2019
3. ISO 3002-3, Basic quantities in cutting and grinding – Part 3: Geometric and kinematic quantities in cutting
4. Nasir, V., Cool, J.: Optimal power consumption and surface quality in the circular sawing process of Douglas-fir wood. Eur. J. Wood Prod. **77**(4), 609–617 (2019). https://doi.org/10.1007/s00107-019-01412-z
5. Mohammadpanah, A., Hutton, S.G.: Dynamic response of guided spline circular saws vs. collared circular saws, subjected to external loads. Wood Mater. Sci. Eng. (2019) https://doi.org/10.1080/17480272.2019.1644371
6. PIŁY LL CUT LINE 3GE I 3GS DO WYCINANIA WAD. http://www.globus-wapienica.eu/podzial_ze_wzgledu_na_narzedzia.25.336.pily_ll_cut_line_3ge_i_3gs_do_wycinania_wad.16.html#.XNQb-hQzaUk Accessed 9 May 2019
7. Chuchala, D., Orlowski, K., Dudek, P.: The methodology for determining of the value of cutting power for cross cutting on optimizing sawing machine. Ann. WULS For. Wood Technol. **103**, 106–113 (2018)
8. Manžos, F.M.: Derevorežušie Stanki, (In Russian: Wood cutting machine tools). Izdatel'stvo Lesnaâ promyšlennost', Moskva (1974)
9. Siklienka, M., Šustek, J., Kminiak, R., Jankech, A.: Delenie a obrábanie dreva. Vysokoškolská učebnica, Technická univerzita vo Zvolene, Zvolen (2017)
10. Svoreň, J., Hrčková, M.: Woodworking machines, Part I. Technical University in Zvolen, Zvolen (2015)
11. http://rema-sa.pl/produkty/optymalizerka-ap-500. Acessed 18 May 2019
12. http://salvadormachines.com/en/supercut-300. Acessed 18 May 2019
13. Astakhov, V.P.: Basic definitions and cutting tool geometry, single point cutting tools. In: Geometry of Single-point Turning Tools and Drills. Fundamentals and Practical Applications. Springer Series in Advanced Manufacturing, pp. 54–101. Springer, London (2010). http://link.springer.com/chapter/10.1007%2F978-1-84996-053-3_2. (Access to this content was enabled by the Library of Gdansk University of Technology. Accessed 31 July 2019)

Wear Characteristics of PA6G Polymer Composite with Oil at Ambient and Elevated Temperatures

Marcin Barszcz[1], Krzysztof Dziedzic[1], Mykhaylo Pashechko[2], and Jerzy Józwik[3](✉)

[1] Department of Computer Science, Electrical Engineering and Computer Science Faculty, Lublin University of Technology, 36B Nadbystrzycka Street, 20-618 Lublin, Poland
[2] Department of Fundamental of Technology, Fundamentals of Technology Faculty, Lublin University of Technology, 38 Nadbystrzycka Street, 20-618 Lublin, Poland
[3] Department of Production Engineering, Mechanical Engineering Faculty, Lublin University of Technology, 36 Nadbystrzycka Street, 20-618 Lublin, Poland
j.jozwik@pollub.pl

Abstract. The article concerns the comparison of tribological properties of polymer composite PA6G with oil at ambient and elevated temperatures. Tribological tests were carried out at ambient temperature and elevated temperatures of 80 °C and 120 °C. The study used a high-temperature tribotester THT 1000 Anton Parr. The tests were conducted in accordance with ASTM G133 and G-99. The friction node was ball-on-disk. Al_2O_3 balls with a diameter of 6 mm were used as counter bodies. The rotating samples were made with polymer composite PA6G oil and had the shape of a disc. Tribological tests were carried out using 10 N load. The speed of sliding friction pair was 0.8 m/s. The friction radius was constant and was 17 mm for all tests. The length of the friction path has been set at 1000 m. During the test, the coefficient of friction, temperature in the friction zone and mass loss were recorded. IT tools were used to process the research results. Multi-criteria evaluation for different operating environments, while taking into account 4 criteria, showed that the work of the material tested at 120 °C is the least favourable.

Keywords: Friction · Polymer composite · PA6G with oil · Elevated temperature

1 Introduction

Currently on the market there is a large amount of materials available for designers and constructors. Therefore, the process of choosing the right materials for specific applications is a difficult challenge for engineers. However, if it is properly made, it can be the basis for long-term work of cooperating elements in the process of their operation [1–3].

In recent years, polymers and polymer composites have been increasingly used in engineering as engineering materials for cooperating machine components [4]. This is influenced by good mechanical and tribological properties [5, 6], resistance to corrosion,

G. M. Królczyk et al. (Eds.): IMM 2019, LNME, pp. 268–278, 2020.
https://doi.org/10.1007/978-3-030-49910-5_24

low weight, flexibility of their processing and self-lubrication, etc. [7, 8]. Thanks to these features, they become alternative materials to traditional metal and ceramic materials. Among the polymer composites, friction-cooperating machine elements are very often used as cast polyamide 6 or PA6C and PA6G. It is dictated by the combination of good mechanical, sliding and wear resistance [5], as well as the possibility of obtaining larger sizes in the casting process. High fatigue strength, noise suppression, corrosion resistance and low weight make it ideal for applications such as rolling and sliding bearings, gears, wedge sheaves and chain sprocket teeth. Thanks to the use of additives such as oil, solid lubricants and thermostabilisers, the typical features of polyamide 6 can be selectively adapted to specific applications, opening up a whole range of possibilities for individual matching of materials for specific purposes. More and more scientists are investigating the tribological properties of polyamide 6 [9–14]. However, research in this area is still very limited. In addition, producers of polymers and polymer composites do not present the results of tribological tests of the materials they produce. In cases where the manufacturer provides tribological data for his materials, they may be the result of limited tests and do not necessarily reflect the actual use [5].

Tribological tests of polyamide 6 were carried out by the authors of the paper [15]. In the article, the effect of carbon fiber (CF) as filler on the tribological properties of the polyamide6/polyphenylene sulfide (PA6/PPS) composites were investigated carefully in order to provide a practical guidance for the use of the polymer-based composites. The paper [16] focuses on exploring the tribological performance of PA-6 reinforced with non-functionalized as well as microwave-functionalized MWCNTs. The authors of the work suggest that this will aid in understanding the polymeric systems that are actually used for heavy duty applications demanding high strength and wear resistance. In the work [17] examined the tribological behavior of polymeric (PTFE, POM-H, PAG6+oil, PA6G and PA6G/Mg) materials during the sliding friction. PA6G+oil material, because of the greasing effect it consists, was observed to have more appropriate wear rates than PA6G and PA6G-Mg [18]. Feyzullahoglu and Saffak [17] also studied the tribological behaviors of different polymer which are among others 6 PLA (Cast Polyamide). Neis et al. [5] investigated of tribological properties of 3 commercially available cast polyamide 6 (PA 6): a natural PA 6 polymer, a PA 6 filled with molybdenum disulfide (MoS_2), and a PA 6 filled with a special solid lubricant. Their tribological results were compared with those measured for a commercial bronze alloy. PA 6 filled with special solid lubricant revealed superior tribological properties among the plastics. In a situation where as a result of the experiment many values are obtained, expressed in different units, informing about the properties of the material tested in various operating conditions, a good approach is to use a multi-criteria assessment. In [19, 20], multi-criteria evaluation was described and applied to issues related to the wear process.

On the basis of the literature analysis, it can be noticed that the conducted research is focused on the effect of additives on tribological properties of polyamide 6. However, there is little tribological testing of PA6G polymer composites with oil and virtually no elevated temperatures at all. In connection with the above, within the framework of this work, tribological tests were performed in commercially available PA6G cast polymer composites with oil at room temperature and elevated temperature.

2 Materials and Methods

The study tested commercially available polymer composites PA6G with oil. According to the manufacturer, PA6G with oil is a self-lubricating polyamide grade designed mainly for the production of friction-cooperating components. The addition of oil causes a significant reduction in the coefficient of friction in relation to the basic form of cast polyamide. The addition of oil also significantly increases the abrasion resistance. It has been developed for non-lubricating moving parts of machines. It comes in three colours: white, yellow and black. Yellow was used for the study [21].

The material for the samples was provided in the form of a sleeve, which was processed by machining (turning and grinding). The surface of the samples was ground with different gradation of grain until the surface roughness was 0.05 μm. The geometry of the samples and their appearance are graphically shown in Fig. 1.

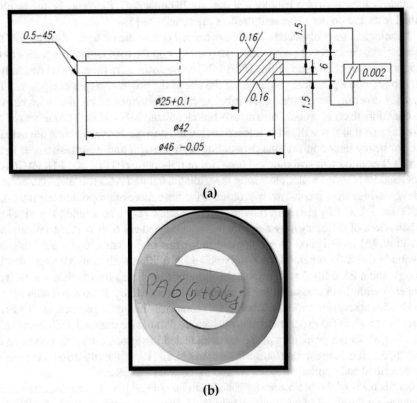

(a)

(b)

Fig. 1. View of samples used for tribological tests: drawing with dimensions (a), picture (b).

Tribological tests were conducted on a high-temperature tribometer, Anton Parr THT 1000, shown in Fig. 2a. The tests were conducted in accordance with ASTM G133 and G-99. The ball-on-disc configuration was used, Fig. 2b. Samples had the shape of a disc. Al_2O_3 balls with a diameter of 6 mm were used as counter bodies. In the tribological tests, loads of 10 N were applied. The tests were conducted with a sliding speed of 0.8 m/s and a radius of 17 mm. The total sliding distance was set equal at 1000 m. Tribological tests were carried out at ambient temperature and elevated temperatures of 80 °C and 120 °C. Temperature was measured with a thermocouple at 2 mm from the surface of friction (actual temperature of the interface and the sample could reach higher values). Variations in the coefficient of friction, friction and temperature were measured in the tests. The frequency of data collection was 10 Hz.

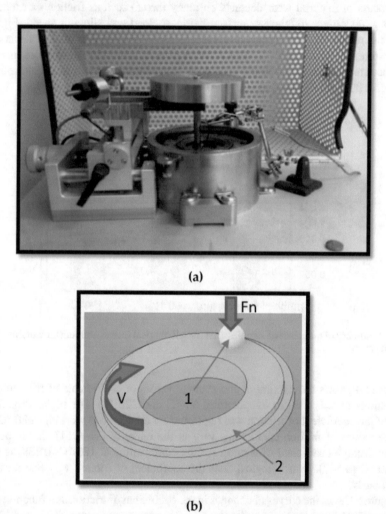

(a)

(b)

Fig. 2. Schematic set-up of tribological tests, THT 1000 tribometer (a), ball-on-disc friction pair (b); 1 – counterbody; 2 – sample.

The surfaces of the samples were subjected to additional tests. The measurements of roughness were conducted and the microscopic traces of friction. Measurement of surface roughness (before and after tribological tests) was made using a needle profilometer

Surtronic 3 by Taylor Hobson. For the evaluation traces of friction, the Alicona metallo-graphic microscope was used. The images were obtained in a digital form and processed using IT tools.

3   Results and Discussion

The process of material wear depends on many factors such as friction pair material, contact temperature, roughness, surface hardness, load and slipping speed. Figure 3 shows the results of measurements of average mass consumption of samples at room temperature, 80 °C and 120 °C. The value of mass consumption of samples increases with the temperature at which the tests were carried out. At room temperature it was 0.005 g, at 80 °C–0.01 g and at 120 °C–0.016 g.

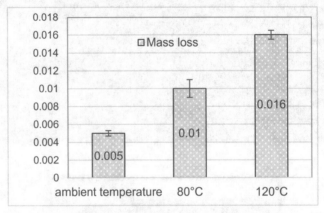

Fig. 3. Summary of mass loss of samples after the tribological tests at ambient temperature, 80 °C and 120 °C.

Figure 4 shows the maximum and average values of coefficients of friction of the associations tested at room temperature, 80 °C and 120 °C. The highest maximum value of the coefficient of friction was registered for associations working at 80 °C. The average values of friction coefficients vary in the range of 0.1–0.147. It can be seen that the average coefficient of friction for the combination at 120 °C (0.139) is lower than 80 °C (0.147). In tribological tests the coefficient of friction was also measured continuously.

Figure 5 shows the curves of changes in the coefficient of friction as a function of the friction path (100 m) for the associations tested at room temperature, 80 °C and 120 °C.

Analysing the results presented in Fig. 5, it can be noticed that the curves show the varied nature of changes. There is also a discrepancy between the curve representing the change in coefficient of friction for the combination working at room temperature and the other curves representing the change in coefficient of friction at 80 °C and 120 °C.

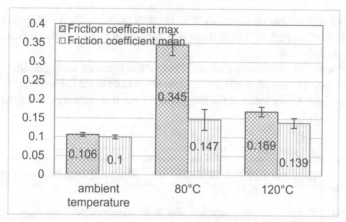

Fig. 4. Comparison of the maximum and average values of coefficients of friction of the tested associations at ambient temperature, 80 °C and 120 °C

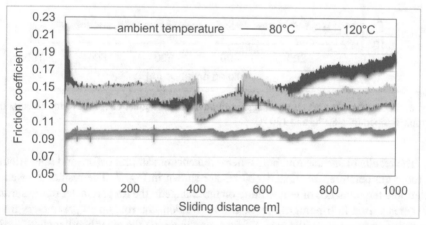

Fig. 5. Changes in the coefficient of friction for the tested associations at room temperature, 80 ° C and 120 °C.

The lowest value of the coefficient of friction was recorded for the combination working at room temperature. The nature of changes in the coefficient of friction is very similar to the linear waveform. It can therefore be stated with some simplification that there is a proportional relationship between the coefficient of friction and the friction. This change is conditioned by wear in the friction node and is almost linear in nature. Noteworthy is the course of the coefficient of friction for the combination working at 120 °C. Its course is different in different friction path ranges. After passing the friction path of 350 m, the first decrease in the friction coefficient takes place. Another rapid drop occurs after reaching the distance of 480 m. On the other hand, after travelling 660 m, there is a rapid increase in value, but then it decreases and stabilises. In [22] this is explained as the result of friction heating, which increases the surface temperature of the contact, leading to the relaxation of polymer molecule chains. A similar course of changes in the coefficient of friction has an association working at 80 °C, with the difference that after crossing the friction path of 350 m there is a rapid drop and from that moment it grows until the end of the test.

The value of the temperature around the friction junction during testing at ambient temperature was almost constant throughout the whole period. However, when tested at 80 °C and 120 °C, its value at the beginning of the test increased and then stabilised (Fig. 6).

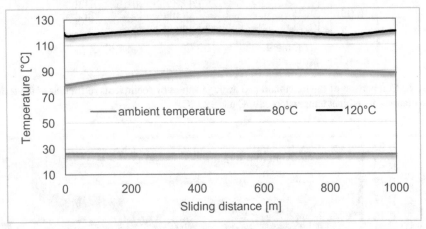

Fig. 6. The course of temperature changes around the friction junction for the tested associations at room temperature, 80 °C and 120 °C

The results of surface roughness measurements of samples before and after friction at room temperature, 80 °C and 120 °C are shown in Fig. 7. The results allowed to determine the influence of temperature on the change in the Ra parameter of the surface of samples used in tribological tests. Before friction, the roughness parameters Ra of the sample surface were 0.15 μm. Surface roughness, on the other hand, increases with increasing temperature, similar to mass consumption. In order to identify the phenomena occurring and the wear mechanism, the friction surfaces were subjected to microscopic observations. Figure 8 shows a view of friction surfaces at room temperature, 80 °C and 120 °C. No significant signs of friction were observed for the surface of samples operating at ambient temperature, Fig. 8a. On the surface, scratches and furrows were observed suggesting micromachining, Fig. 8b, c. The depth of cut increased with increasing temperature.

From the point of view of the exploitation of the material tested, the issue of wear at elevated temperatures is of interest. It is obvious that at ambient temperature consumption will be significantly lower. The multi-criteria assessment of the effects of consumption was based on four criteria, all of which were minimised. Criterion 1 – mass loss, Criterion 2 – friction coefficient max., Criterion 3 – friction coefficient mean, Criterion 4 – Ra roughness increase after testing. In the first stage, a subset of non-dominated solutions

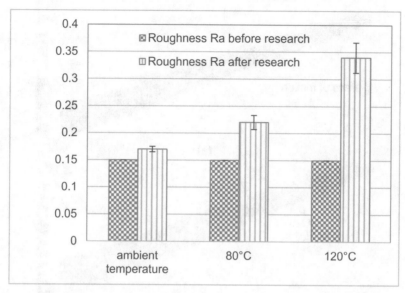

Fig. 7. Results of surface roughness measurements of samples before and after friction, at room temperature, 80 °C and 120 °C

was determined and then, using the Chebyshev metrics, a min-max method was built using weights, which allowed to determine a compromise solution.

$$p[F(x)] = \min_{n \in N} \max_{j \in J} \left\{ \omega_j \frac{\left| F_j^0 - F_j^n \right|}{F_j^0} \right\} \tag{1}$$

ω_j – "weight" of the j-th criterion, the sum of all weights equals 1,
j, n – indexes of the criteria and non-dominant solutions considered,
F_{oj} – j-th component of the ideal vector.

After omitting the values determined at the ambient temperature, it turns out that the results achieved at temperatures of 80 °C and 120 °C create optimal solutions in the Pareto sense. For higher temperatures, lower values of Criterion 2 and 3 were obtained than for the temperature of 80 °C. The search for solutions with the min-max method with weights is presented in Table 1.

The results obtained show that only very strong preference for Criterion 3 (3 times more than Criterion 1 and 6 times more than Criterion 4) leads to selection of the 120 °C environment as more suitable for exploiting the coating tested.

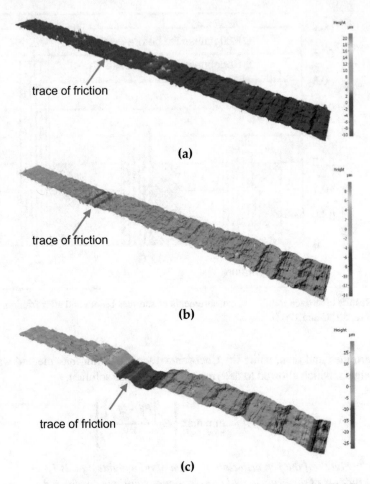

Fig. 8. View of the surface of samples after friction at: room temperature (a), 80 °C and 120 °C (b, c).

Table 1. The search for solutions with the min-max method with weights

Lp.	Weight	The compromise solution
1.	0.25 0.25 0.25 0.25	at 80°
2.	0.20 0.20 0.40 0.20	at 80°
3.	0.20 0.10 0.60 0.10	at 120°

4   Conclusion

The tests show that PA6G with oil in cooperation with Al_2O_3 aluminum oxide shows good tribological properties not only at ambient temperature but also at elevated temperatures of 80 °C and 120 °C. This is confirmed by low values of mass consumption and friction coefficients. Similar results were obtained in [23], where different polymer composites were tested at room temperature at 10, 20 and 30 N load (it was shown that the PA6-G composite with mineral oil obtained the best results among all tested polymer composites).

The friction coefficient values obtained between 0.1–0.147 are a very good result, considering that the composites were loaded with the concentrated force exerted by the hard material (Al_2O_3 aluminum oxide). It was observed that the average coefficient of friction for the combination working at 120 °C is lower (0.139) than at 80 °C (0.147). It was also noticed that mass consumption increases with the temperature of the tests being carried out. However, the mass consumption results obtained at 120 °C are satisfactory (0.016 g). Taking into account the four-element set of criteria, the multi-criteria analysis showed that the most unfavourable environment for exploiting the material tested is 120 °C, although Criteria 2 and 3 had values lower than at 80 °C. It can therefore be concluded that composites are able to transfer large loads exerted by much harder material, while maintaining very good tribological properties at the level of associations of even such materials as steel-graphite or steel-glass.

Acknowledgments. The project/research was financed in the framework of the project Lublin University of Technology - Regional Excellence Initiative, funded by the Polish Ministry of Science and Higher Education (contract no. 030/RID/2018/19).

References

1. Keresztes, R., Odrobina, M., Nagarajan, R., Subramanian, K., Kalacska, G., Sukumaran, J.: Tribological characteristics of cast polyamide 6 (PA6G) matrix and their composite (PA6G SL) under normal and overload conditions using dynamic pin-on-plate system. Compos. Part B Eng. **160**, 119–130 (2019)
2. Maruda, R.W., Feldshtein, E., Legutko, S., Królczyk, G.M.: Improving the efficiency of running-in for a bronze-stainless steel friction pair. J. Frict. Wear. **36**(6), 548–553 (2015)
3. Maruda, R.W., Feldshtein, E., Legutko, S., Królczyk, G.M.: Analysis of contact phenomena and heat exchange in the cutting zone under minimum quantity cooling lubrication conditions. Arab. J. Sci. Eng. **41**(2), 661–668 (2016)
4. Friedrich, K.: Polymer composites for tribological applications. Adv. Ind. Eng. Polym. Res. **1**, 3–39 (2018)
5. Neis, P.D., Ferreira, N.F., Poletto, J.C., Sukumaran, J., Andó, M., Zhang, Y.: Tribological behavior of polyamide-6 plastics and their potential use in industrial applications. Wear **376–377**, 1391–1398 (2017)
6. Song, F., Wang, Q., Wang, T.: The effects of crystallinity on the mechanical properties and the limiting PV (pressure × velocity) value of PTFE. Tribol. Int. **93**, 1–10 (2015)
7. Liu, Y., et al.: Preparation and tribological properties of hybrid PTFE/Kevlar fabric self-lubricating composites. Surf. Coat. Technol. **361**, 196–205 (2019)

8. Li, H., Li, S., Li, F., Li, Z., Wang, H.: Fabrication of SiO_2 wrapped polystyrene microcapsules by Pickering polymerization for self-lubricating coatings. J. Colloid Interface Sci. **528**, 92–99 (2018)

9. Karami, P., Shojaei, A.: Improvement of dry sliding tribological properties of polyamide 6 using diamond nanoparticles. Tribol. Int. **115**, 370–377 (2017)

10. Karatas, E., Gul, O., Karsli, N.G., Yilmaz, T.: Synergetic effect of graphene nanoplatelet, carbon fiber and coupling agent addition on the tribological, mechanical and thermal properties of polyamide 6,6 composites. Comp. Part B Eng. **163**, 730–739 (2019)

11. Zhou, S., Wang, J., Wang, S., Ma, X., Huang, J., Zhao, G., Liu, Y.: Facile preparation of multiscale graphene-basalt fiber reinforcements and their enhanced mechanical and tribological properties for polyamide 6 composites. Mater. Chem. Phys. **217**, 315–322 (2018)

12. You, Y.-L., Li, D.-X., Si, G.-J., Deng, X.: Investigation of the influence of solid lubricants on the tribological properties of polyamide 6 nanocomposite. Wear **311**, 57–64 (2014)

13. Zhao, L.-x., Zheng, L.-y., Zhao, S.-g.: Tribological performance of nano-Al_2O_3 reinforced polyamide 6 composites. Mater. Lett. **60**, 2590–2593 (2006)

14. Li, D.-X., Deng, X., Wang, J., Yang, J., Li, X.: Mechanical and tribological properties of polyamide 6 polyurethane block copolymer reinforced with short glass fibers. Wear **269**, 262–268 (2010)

15. Zhou, S., Zhang, Q., Wu, Ch., Huang, J.: Effect of carbon fiber reinforcement on the mechanical and tribological properties of polyamide6/polyphenylene sulfide composites. Mater. Des. **44**, 493–499 (2013)

16. Chopra, S., Deshmukh, K.A., Deshmukh, A.D., Peshwe, D.R.: Functionalization and melt-compounding of MWCNTs in PA-6 for tribological applications. IOP Conf. Ser.: Mater. Sci. Eng. **346**, 1–8 (2018)

17. Zsidai, L., De Baets, P., Samyn, P., Kalacska, G., Van Peteghem, A.P., Van Parrys, F.: The tribological behaviour of engineering plastics during siliding friction investigated with small scale specimens. Wear **253**, 673–688 (2002)

18. Feyzullahoglu, E., Saffak, Z.: The tribological behaviour of different engineering plastics under dry friction conditions. Mate. Des. **29**, 205–211 (2008)

19. Montusiewicz, J., Osyczka, A.: A decomposition strategy for multicriteria optimization with application to machine tool design. Eng. Costs Prod. Econ. **20**, 191–202 (1990)

20. Pashechko, M., Montusiewicz, J., Dziedzic, K., et al.: Multicriterion assessment of wear resistance of Fe–Mn–C–B eutectic coatings alloyed with Si, Ni, and Cr. Powder Metall. Met. Ceram. **56**, 316–322 (2017)

21. https://www.ensingerplastics.com

22. Unal, H., Mimaroglu, A.: Friction and wear behaviour of unfilled engineering thermoplastics. Mater. Des. **24**(3), 183–187 (2003)

23. Józwik, J., Kieliszek, K.: Ocena porównawcza właściwości tribologicznych wybranych polimerów i kompozytów polimerowych. Materiały Kompozytowe **2**, 55–60 (2018)

Analysis of the Friction and Wear Mechanisms of AlTiN Coated Carbide in Combination with Inconel 718 Nickel Alloy

Marta Bogdan-Chudy and Piotr Niesłony[✉]

Department of Manufacturing Engineering and Automation, Opole University of Technology,
5 Mikołajczyka Str., 47-271 Opole, Poland
p.nieslony@po.edu.pl

Abstract. The paper presents the results of tribological examination of Inconel 718 nickel alloy in combination with sintered carbide with PVD-AlTiN coating under dry friction conditions. Experimental studies were conducted on a pin-on-disc tribometer, which enables continuous recording of dynamic changes in the normal force F_N and friction force values. On this basis, both instantaneous and averaged values of the coefficient of friction μ as a function of the friction path were determined. The studies were conducted for sliding speed $v_s = 50 \div 200$ m/min and for $F_N = 60$ and 120 N. The influence of slip velocity and normal force on the changes of μ value and the effects of wear of friction pairs were evaluated. Scanning electron microscopy (SEM), EDS analysis and confocal and optical microscopy were used to identify wear mechanisms. The analysis of surface wear and tear of samples and counter-samples was made by examining the friction areas and mechanisms of material transfer within the friction pairs. A significant influence of slip velocity on the decrease of friction coefficient value and stabilization of friction conditions in tribological pair was found.

1 Introduction

Frictional studies conducted in the pin-on-disc system provide reliable information on tribological characteristics of the cutting process. In the literature on the subject there are many papers on friction of difficult-to-machine materials. As part of the work [1–6], investigations were carried out into the friction of titanium alloys and nickel alloys with sintered carbide tools [1, 6, 7]. The work by Courbon et al. is particularly interesting, [1] where detailed investigations of friction of Inconel 718 nickel alloy in combination with a TiN coated sintered carbide pin and Ti6Al4V titanium alloy in combination with uncoated sintered carbide were carried out. In the study, not only the value of the friction coefficient was analyzed, but also the wear and tear of the pin surface.

On the basis of analyzed studies it was found that the greatest influence on the average friction coefficient value is that of the slip velocity at which materials move relative to each other in a friction pair [1, 7–9]. In most of the analyzed publications the authors evaluated the changes in the friction coefficient. Only few studies cover comprehensive assessment of tribological interactions, e.g. through an analysis of phenomena of wear and tear of tribological pair materials.

© Springer Nature Switzerland AG 2020
G. M. Królczyk et al. (Eds.): IMM 2019, LNME, pp. 279–291, 2020.
https://doi.org/10.1007/978-3-030-49910-5_25

Nickel-based alloys belong to the group of heat-resistant and heat-temperature creep resistant materials (HRSA) [Heat Resistant Super Alloys] most commonly used in the aviation industry, including heat-stressed turbine bodies, combustion chambers, exhaust valves, exhaust ducts, discs, shafts, turbine blades and compressor blades, etc.

The problems of poor machinability of nickel alloys result from the mechanical properties of these materials, which include high creep resistance and increased resistance to thermal and mechanical shocks [10]. The low thermal conductivity of nickel alloys ($11.4 \text{ W/m}^2\text{K}$) leads to high temperatures in the cutting zone and heating of the tool [11]. Kitagawa and others [12] showed that the temperature in the cutting zone of Inconel 718 alloy increases to approximately 900 °C at a relatively low cutting speed of 30 m/min, and up to 1300 °C at 300 m/min. The presence of large amounts of elements such as nickel, chromium, cobalt and titanium is the reason for poor machinability of high temperature super alloys. Most of the problems during the machining of nickel alloys are caused by heat generation, mainly during its decohesion as well as friction in the tool-chip and tool-surface arrangement [13]. The heat generated during the machining of nickel alloys affects the microstructure of the alloy and causes the formation of increased thermal stresses, which translates into durability, fatigue resistance, performance and functionality of the manufactured object.

Since nickel alloy 718 belongs to the group of difficult to machine materials, therefore, the knowledge of tribological phenomena occurring at the contact of the friction pair of nickel alloy Inconel 718 - sintered carbide is very valuable. The knowledge gained in this way can be used, among other things, for comprehensive modeling of the cutting process by FEM [Finite Element Method] techniques.

For this reason, the results of tribological research of the pair of nickel alloy Inconel 718 - sintered carbide covered with AlTiN coating, carried out for a wide range of slip velocities, presented in this article, seem to be the most rational. The information obtained may contribute to a better understanding of tribological phenomena and wear mechanisms occurring in this type of friction pairs.

2   Set up and Experimental Methodology

Tribological research was carried out on an original tribometer of the "pin-on-disc" type (Fig. 1). This stand enables controlled cooperation of tribological pair in the sample-countersample system. The design of the test stand allows for smooth adjustment and stabilization of the normal force F_N and slip velocity v_s. The friction and normal force were measured using a piezoelectric dynamometer Kistler 9129AAA. The signal from the dynamometer was transmitted by the Kistler 5070A10100 amplifier to a computer using a National Instrument measuring card where, in the Labview environment, the signals were conditioned and the data obtained were analyzed.

During the friction tests, the averaged values of the friction coefficient were determined on the basis of measurements from a piezoelectric dynamometer and the dynamics of the process was analyzed on the basis of normal force and friction measurements made with a high sampling frequency of 6000 Hz.

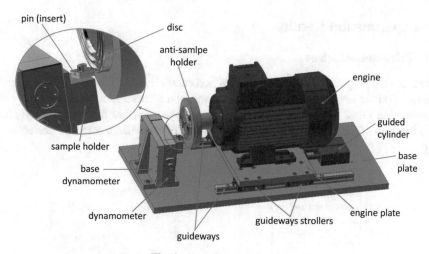

Fig. 1. Experimental set-up

Friction tests were carried out in a system constituting a classical tribological pair of a sample - countersample (Fig. 2a). The anti-sample disc was made of Inconel 718 nickel alloy. Three paths were made on it with variable diameters and a profiled contact surface to ensure proper contact (Fig. 2b).

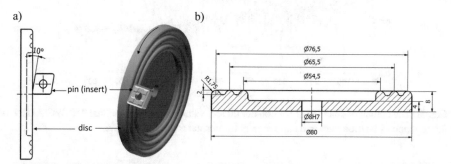

Fig. 2. Configuration of the tribo-pair (a) and profiles of friction paths on the Inconel 718 disc (b).

A sample in friction testing was a commercial cutting insert type CNMG 120412-UP KC5010 by Kennametal, made of sintered carbide with PVD-Al$0_{0.55}$ Ti$_{0.45}$N coating. Tribological investigations were carried out under dry friction conditions in unidirectional motion with a normal force F_N of 60 and 120 N, for friction path $s = 1000$ m and for three selected slip velocities v_s: 50, 100, 150 and 200 m/min.

The HRM-300M Huvitz metallographic microscope, the Olympus IX70 optical microscope and the Sensofar 3D S neox profilometer were used to measure wear and visualize sample surfaces. Metallographic examination was carried out with the use of scanning electron microscope (SEM) type JXA-840A by Joel equipped with the EDS chemical composition analysis system.

3 Experimental Results

3.1 Friction Coefficient

The recorded progress of changes in the friction coefficient as a function of friction path values obtained in tribological tests for 2 normalized forces $F_N = 60$ N and 120 N are presented in Figs. 3 and 4, respectively. The average values of the friction coefficient as a function of the slip velocity for the same cases of normalized force are presented in Fig. 7.

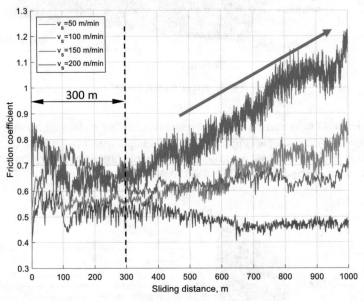

Fig. 3. The course of changes of friction coefficient value for the pair Inconel 718-WC/AlTiN as the function of friction path and slip velocity for normal force $F_N = 60$ N.

The greatest changes in the friction coefficient were observed for both normal force cases $F_N = 60$ and 120 N for the friction path up to 300 m (Fig. 3 and 4). In this respect, there was intensive matching of the elements of the friction pair. Their detailed analysis is presented in Fig. 5 and 6. For $F_N = 60$ N, a monotonic increase of the friction coefficient occurred with the increase of the friction path, especially past 300 m. This is clearly noticeable for the $v_s = 50 \div 150$ m/min (Fig. 3). For $F_N = 120$ N, this trend is much smaller in scale. It seems that with the higher normal force conditions stabilized faster in the friction pair. This can also be confirmed by the fact that this phenomenon is clearly visible for $v_s = 50$ m/min. The lower the slip velocity, the more dominant the adhesive actions become. This is also shown in Sect. 3.2.

Due to the high sampling frequency of the friction and normal force readings, it was possible to record momentary changes in the friction coefficient. It can be stated that the highest fluctuations of momentary μ changes occurred for $F_N = 60$ N and $v_s = 50$ m/min (Fig. 3). This may be a confirmation of the dominant role of adhesion phenomena at

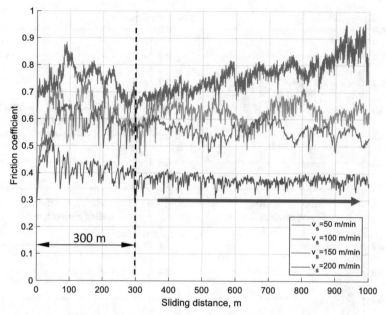

Fig. 4. The course of changes of friction coefficient value for the pair Inconel 718-WC/AlTiN as the function of friction path and slip velocity for normal force $F_N = 120$ N.

the pin-disc contact. The formation and breakage of an adhesive micro-connection may cause such course of μ signal.

It should be noted that for both cases of the determined F_N force, the friction coefficient as a function of the friction path decreases for the highest slip velocity $v_s = 200$ m/min (Figs. 3 and 4). This phenomenon may result from the formation of favorable conditions for slippage in the friction pair (through plasticizing the material as a result of, among other things, the temperature increase). This also has practical implications. It seems that it is possible to find the optimum cutting conditions, where by increasing the cutting speed we reduce the friction on the surface of the tool's rake face (chip slip). Reduced friction results in the reduction of the friction temperature, which is a positive phenomenon expected by the user.

A graphical representation of the changes in friction coefficient values for each slip velocity at the initial stage of tribological pair interaction is shown in Figs. 5 and 6. In all cases, a rapid increase in μ occurs immediately after the start of the test. It has been noted for $F_N = 60$ N (Fig. 5) that the higher the speed v_s, the path necessary to reach this maximum point slightly increases. For $F_N = 120$ N this relationship is not so obvious (Fig. 6). In this case, the fluctuation of μ changes is much higher, which makes it difficult to determine the unequivocal character of the coefficient course in the initial period of friction. It should be noted, however, that in this case the process is carried out at twice the normal force.

The graphs in Fig. 5d and 6d confirm the rapid stabilization of the friction conditions for the highest slip velocity ($v_s = 200$ m/min). In addition, a significant decrease in μ

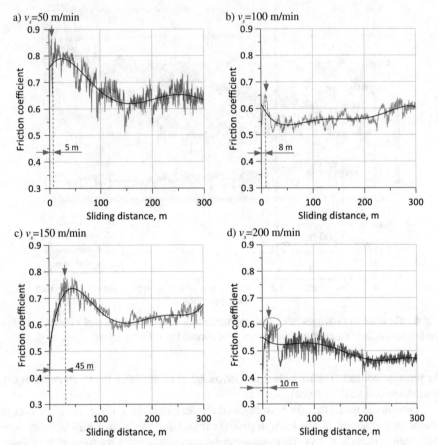

Fig. 5. The course of changes in the Inconel 718-WC/AlTiN friction coefficient for the friction path up to 300 m and $v_s = 50 \div 200$ m/min. Normal force 60 N.

value is also clearly visible already in the initial period of friction. As shown in Fig. 3 and 4, this trend for $v_s = 200$ m/min is noticeable for the whole range of the friction path.

The static evaluation of the friction tests was based on the average coefficient of friction. The mean value was calculated for the whole friction path of 1000 m. The results of these calculations are shown in Fig. 7. In general, the average coefficient of friction decreased with increasing slip velocity. This was observed for both normal forces tested.

For the lowest slip velocity $v_s = 50$ m/min the highest mean value of the coefficient of friction was obtained at the level of $\mu = 0,82 \pm 0,16$ for $F_N = 60$ N and $\mu = 0,76 \pm 0,06$ for $F_N = 120$ N. Additionally, high fluctuations of μ values were confirmed exactly for $v_s = 50$ m/min and $F_N = 60$ N (Fig. 7a).

For the highest test speed the mean value of the coefficient of friction is the lowest and equals $\mu = 0,50 \pm 0,04$ for $F_N = 60$ N and $\mu = 0,38 \pm 0,03$ for $F_N = 120$ N, respectively. In general, lower μ values were obtained for $F_N = 120$ N.

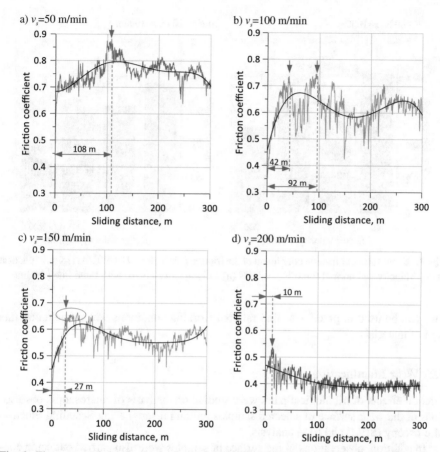

Fig. 6. The course of changes in the Inconel 718-WC/AlTiN friction coefficient for the friction path up to 300 m and $v_s = 50 \div 200$ m/min. Normal force 120 N.

The trend of changes in the average values of the coefficient of friction assessed for both cases of normal force F_N is similar. In both cases, the linear fitting gives a good approximation. For $F_N = 60$ N the trend line is described by the equation $\mu = 0.001908 * v_s + 0.8891$ with a fitting of R-squared $= 0.869$. For $F_N = 120$ N the trend line is described by the equation $\mu = -0.0023468 * v_s + 0.87555$ with R-squared $= 0.965$. Based on these equations, it was calculated that for $F_N = 60$ N the mean gradient reduction with an increase in v_s equals 0.0019 [m/min]$^{-1}$ (Fig. 7a) where for $F_N = 120$ N it is 0.0023 [m/min]$^{-1}$ (Fig. 7b).

It is interesting that in the investigated case a twice higher normal force caused a linear decrease in the mean coefficient of friction from about 0.035 for $v_s = 50$ m/min down to 0.101 for $v_s = 200$ m/min.

These tests show that an increase in the pressure force of the specimen against the rotating disc may, in the range of technological parameters studied, reduce the friction coefficient, and thus reduce the wear of the cooperating elements of the friction pair.

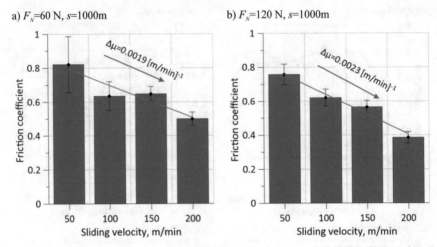

Fig. 7. Mean values of friction coefficient of the friction pair Inconel 718-WC/AlTiN as a function of slip velocity and normal force $F_N = 60$ N (a) and $F_N = 120$ N (b) with trend line (red line)

This can be used in practice to find rational working conditions when machining this type of materials.

3.2 Wear Identification

Friction effects of tribological pairs were studied on the basis of microscopic observations of the wear surface of selected samples (cutting inserts) using Scanning Electron Microscopy (SEM) and EDS analysis.

In addition, observations of the surface of samples were also carried out using confocal and optical microscopy. The application of various methods of observation and analysis of the surface of the tested materials allowed to present the effects and phenomena of wear in various aspects. Analysis of the chemical composition of EDS and measurement on confocal microscopy allowed to confirm the observed wear and tear results obtained with optical microscopy. Surface profiles of abraded samples, obtained on the basis of confocal microscopy, were developed in MountainsMap Premium 2.4 software from Digital Surf.

Figures 8 and 9 show examples of surface friction images of WC/AlTiN samples made with different imaging techniques.

It was observed that the nature of the surface wear is comparable for all samples. The AlTiN coating has been partially removed in many places where the sample has come into contact with the countersample. In the whole range of examined parameters, adhesive accretions and residues of transferred countersample material from Inconel 718 nickel alloy were found. The results of EDS chemical composition analysis (Fig. 10) allow to state that during friction there was a mutual movement of materials and their deposition in the contact area.

In the contact areas on the surface of the sample, build-ups of the countersample material are formed, which may indicate a high contribution of adhesion in the complex

Fig. 8. SEM photo of WC/AlTiN sample surface area after friction tests of the pair Inconel 718-WC/AlTiN for $F_N = 60$ N, $v_s = 50$ m/min, $s = 1000$ m.

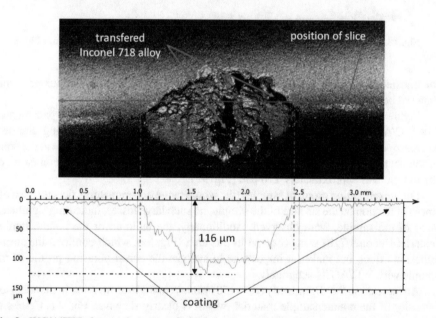

Fig. 9. WC/AlTiN plate surface area after friction tests of the pair Inconel 718-WC/AlTiN for $F_N = 60$ N, $v_s = 100$ m/min, $s = 1000$ m. Surface stereometric view (a), surface profile at the cross-sectional area (b)

mechanism of wear and tear of the tested samples. The AlTiN coating is subject to systematic wear and tear, its losses are visible (point #2 in Fig. 8) as well as adhesive sticking of the countersample material are visible in the form of a discontinuous layer of

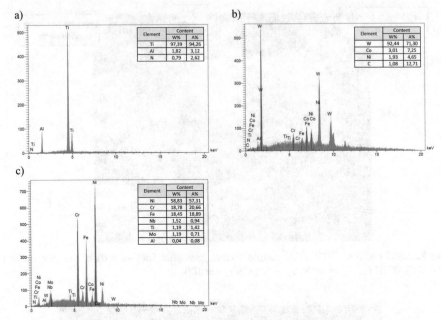

Fig. 10. EDS spectrum at the point: #1 (a), #2 (b) and #3 (c) for the sample from Fig. 8

the transferred material (point #3 in Fig. 8). The area of the transferred countersample material becomes wider as the slip speed increases.

Figure 9 shows an image of WC/AlTiN plate after friction tests of the pair Inconel 718-WC/AlTiN for $F_N = 60$ N, $v_s = 100$ m/min and $s = 1000$ m using confocal microscope. Based on this image, MountainsMap generated a surface profile at the point of the deepest wear. It was measured that in this case the maximum abrasion depth of the sample was approximately 116 μm (Fig. 9b).

For the remaining group of samples (the whole range of parameters), despite the clear traces of friction on the surface of the sample, no such large losses, indicating significant wear of the sample, were observed. Additionally, the build-up of the countersample material (Inconel 718) was observed in this area (Fig. 9a), which confirms the strong adhesion effect, but makes it impossible to analyze the actual abrasion profile of the sample with WC/AlTiN accurately.

An increase in the slip velocity ($v_s = 150$ m/min) results in an increase in the adhesion intensity of the countersample material, which is clearly shown in Fig. 11. Cracks in the deposited material have been observed, which may indicate its multilayer character, probably resulting from its multiple application to the same place.

Friction effects were also studied on the second element of the tribological pair - the countersample made of Inconel 718 nickel alloy. Photographs taken with the optical microscope are shown in Figs. 12 and 13.

In principle, friction effects are comparable for the entire range of parameters tested. Only for the lowest slip velocity $v_s = 50$ m/min at $F_N = 60$ N the grinding of the anti-sample material was observed (Fig. 12a). Also in this case, the coefficient of friction

Fig. 11. Optical microscope image of WC/AlTiN plate surface after Inconel 718-WC/AlTiN pair friction tests for $F_N = 60$ N, $v_s = 150$ m/min, $s = 1000$ m

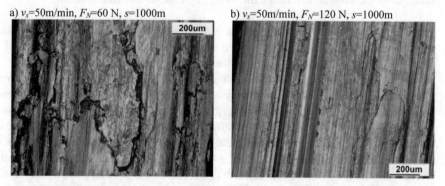

Fig. 12. Images of the surface of countersamples of Inconel 718 alloy after friction tests with WC/AlTiN sintered carbide sample at two values of normal force and path and at constant v_s.

was characterized by the highest fluctuations. The lack of homogeneous slip path and intensive grinding of the material caused by the plasticization of the top contact layer of Inconel could have caused such intense momentary changes in the friction coefficient observed in the course in Fig. 3.

For higher friction speeds, regardless of the normal force, only slight grooving marks are visible on the friction paths (Fig. 13). These are typical effects of adjusting the countersample material to the friction conditions in a pin-on-disc system.

a) v_s=150m/min, F_N=60 N, s=1000m   b) v_s=150m/min, F_N=120 N, s=1000m

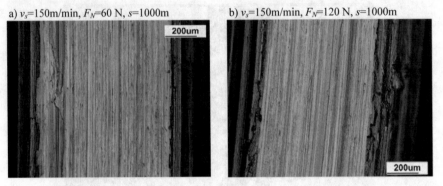

Fig. 13. Images of the surface of countersamples of Inconel 718 alloy after friction tests with WC/AlTiN sintered carbide specimens at two values of F_N normal and constant force v_s.

4 Conclusions

Based on the data obtained from tribological research in the "pin-on-disc" system conducted on the original test stand, the following conclusions can be drawn:

– It was observed that the average values of the coefficient of friction decrease with increasing slip velocity, and the gradient of the decrease of μ as a function of v_s is relatively constant for individual F_N values.
– Under technically dry friction conditions, the lowest friction coefficient values for Inconel 718-WC/AlTiN pair were obtained for the highest tested slip velocity $v_s =$ 200 m/min, regardless of the applied normal force. In addition, it was observed that the value of μ decreases with the increase of the friction path. Reduced friction results in the reduction of the friction temperature, which is a positive phenomenon expected by the user.
– When evaluating the course of μ changes as the function of friction path, it was found that at the beginning of the process there was a rapid increase in its value to the maximum values for the whole course. The distance of the friction path necessary to reach the maximum point slightly increases with increasing v_s, however, this phenomenon is pronounced only for $F_N = 60$ N.
– It was observed that in the whole range of tested friction parameters the character of surface wear of samples is comparable. In many places the AlTiN coating was partially removed and Inconel 718 adhesive build-ups appeared.
– Similarly, the friction effects observed on the countersample of Inconel 718 are comparable for the whole range of tested parameters, and only traces of material grinding were observed for $v_s = 50$ m/min at $F_N = 60$ N.

To sum up, it can be stated that tribological investigations using tribometers allow for modelling and experimental evaluation of hard-to-verify tribological phenomena occurring at the tool-chip contact during the actual cutting process. In the investigated case, the use of higher friction speeds, and thus higher cutting speeds, reduces the friction coefficient of the chip against the surface of the sintered carbide cutting inserts, thus reducing one of the factors influencing the durability of the tool and its friction wear.

The knowledge of the process of tool-chip friction cannot be overestimated in industrial practice because it allows for rationalization of friction conditions, mainly in terms of maximizing the tool life of the cutting insert.

This knowledge is also important for the numerical modelling of the cutting process. Information about the friction conditions at the tool-chip interface is one of the important factors influencing the reliability of the simulation data obtained.

References

1. Courbon, C., Pusavec, F., Dumont, F., Rech, J., Kopac, J.: Tribological behaviour of Ti6Al4V and Inconel718 under dry and cryogenic conditions - application to the context of machining with carbide tool. Tribolology Int. **66**, 72–82 (2013)
2. Egana, A., Rech, J., Arrazola, P.J.: Characterization of friction and heat partition coefficients during machining of a TiAl6V4 titanium alloy and a cemented carbide. Tribology Trans. **55**, 665–676 (2012)
3. Ezugwu, E.O., Bonney, J., Fadare, D.A., Sales, W.F.: Machining of nickel-base, Inconel 718, alloy with ceramic tools under finishing conditions with various coolant supply pressures. J. Mater. Process. Technol. **162–163**, 609–614 (2005)
4. Grzesik, W., Małecka, J., Zalisz, Z., Żak, K., Niesłony, P.: Investigation of friction and wear mechanisms of AlTiN coated carbide against Ti6Al4V titanium alloy using pin-on-disc tribometer. Arch. Mech. Eng. **LXIII**, 113–127 (2016)
5. Smolenicki, D., Boos, J., Kuster, F., Roelofs, H., Wyen, C.F.: In-process measurement of friction coefficient in orthogonal cutting. CIRP. Ann. – Manuf. Technol. **63**(1), 97–100 (2014)
6. Bogdan-Chudy, M., Niesłony, P.: Ocena warunków tribologicznych podczas skrawania stopu Inconel 718 płytką z węglika spiekanego. Mechanik **415**(8–9), 91–97 (2015)
7. Zemzemi, F., Rech, J., Ben Salem, W., Dogui, A., Kapsa, P.: Identification of friction and heat partition model at the tool-chip-workpiece interfaces in dry cutting of an Inconel 718 alloy with CBN and coated carbide tool. Adv. Manuf. Sci. Technol. **38**, 5–22 (2014)
8. Bonnet, C., Valiorgue, F., Rech, J., Hamdi, H.: Improvement of the numerical modeling in orthogonal dry cutting of an AISI 316L stainless steel by the introduction of a new friction model. CIRP J. Manuf. Sci. Technol. **1**(2), 114–118 (2008)
9. Hong, P.P.Y., Ding, Y., Jeong, W.: Friction and cutting forces in cryogenic machining of Ti-6Al-4V. Int. J. Mach. Tools Manuf **41**(15), 2271–2285 (2001)
10. Ulutan, D., Özel, T.: Machining induced surface integrity in titanium and nickel alloys: a review. Int. J. Mach. Tools Manuf **51**(3), 250–280 (2011)
11. Pande, P.P., Sambhe, R.U.: Machinability assessment in turning of Inconel 718 nickel-base super alloys: a review. Int. J. Mech. Eng. Technol. **5**(10), 94–105 (2014)
12. Kitagawa, T., Kubo, A., Maekawa, K.: Temperature and wear of cutting tools in high-speed machining of Inconel 718 and Ti-6Al-6V-2Sn. Wear **202**(2), 142–148 (1997)
13. Ezugwu, E.O.: Key improvements in the machining of difficult–to–cut aerospace superalloys. Int. J. Mach. Tools Manuf **45**(12–13), 1353–1367 (2005)

Author Index

© Springer Nature Switzerland AG 2020
G. M. Królczyk et al. (Eds.): IMM 2019, LNME, pp. 293–294, 2020.
https://doi.org/10.1007/978-3-030-49910-5

Printed in the United States
By Bookmasters